Mathematical Principles of Mechanics and Electromagnetism

Part B: Electromagnetism and Gravitation

MATHEMATICAL CONCEPTS AND METHODS IN SCIENCE AND ENGINEERING

Series Editor: **Angelo Miele**
Mechanical Engineering and Mathematical Sciences
Rice University

Volume 1 **INTRODUCTION TO VECTORS AND TENSORS**
Volume 1: Linear and Multilinear Algebra
Ray M. Bowen and C.-C. Wang

Volume 2 **INTRODUCTION TO VECTORS AND TENSORS**
Volume 2: Vector and Tensor Analysis
Ray M. Bowen and C.-C. Wang

Volume 3 **MULTICRITERIA DECISION MAKING AND DIFFERENTIAL GAMES**
Edited by George Leitmann

Volume 4 **ANALYTICAL DYNAMICS OF DISCRETE SYSTEMS**
Reinhardt M. Rosenberg

Volume 5 **TOPOLOGY AND MAPS**
Taqdir Husain

Volume 6 **REAL AND FUNCTIONAL ANALYSIS**
A. Mukherjea and K. Pothoven

Volume 7 **PRINCIPLES OF OPTIMAL CONTROL THEORY**
R. V. Gamkrelidze

Volume 8 **INTRODUCTION TO THE LAPLACE TRANSFORM**
Peter K. F. Kuhfittig

Volume 9 **MATHEMATICAL LOGIC:** An Introduction to Model Theory
A. H. Lightstone

Volume 11 **INTEGRAL TRANSFORMS IN SCIENCE AND ENGINEERING**
Kurt Bernardo Wolf

Volume 12 **APPLIED MATHEMATICS:** An Intellectual Orientation
Francis J. Murray

Volume 14 **PRINCIPLES AND PROCEDURES OF NUMERICAL ANALYSIS**
Ferenc Szidarovszky and Sidney Yakowitz

Volume 16 **MATHEMATICAL PRINCIPLES OF MECHANICS AND ELECTROMAGNETISM,** Part A: Analytical and Continuum Mechanics
C.-C. Wang

Volume 17 **MATHEMATICAL PRINCIPLES OF MECHANICS AND ELECTROMAGNETISM,** Part B: Electromagnetism and Gravitation
C.-C. Wang

Mathematical Principles of Mechanics and Electromagnetism

Part B: Electromagnetism and Gravitation

C.-C. Wang
Rice University
Houston, Texas

PLENUM PRESS · NEW YORK AND LONDON

Library of Congress Cataloging in Publication Data

Wang, Chao-Cheng, 1938-
Mathematical principles of mechanics and electromagnetism.

(Mathematical concepts and methods in science and engineering; v. 16–17)
Bibliography: pt. A, p. ; pt, B, p.
Includes Indexes.
CONTENTS: pt. A. Analytical and continuum mechanics.–pt. B. Electromagnetism and gravitation.
1. Mechanics, Analytic. 2. Continuum mechanics. 3. Electromagnetism. 4. Gravitation. I. Title.
QA805.W26 531'.0151 79-11862
ISBN 0-306-40212-2 (v. 17)

A Division of Plenum Publishing Corporation
227 West 17th Street, New York, N.Y. 10011

Printed in the United States of America

To my teacher
CLIFFORD TRUESDELL

Preface to Part B

In Part B, I present the mathematical principles of electromagnetism and gravitation, showing how these principles evolve from the classical forms to the relativistic forms.

I start from the concept of the ether frame, which is taken to be a particular inertial frame in the Newtonian space–time. Relative to this frame I formulate the mathematical principles of electrostatics, magnetostatics, time-varying electromagnetic fields, and electromechanical interactions in moving media. These principles give rise to the field equations, known as Maxwell's equations. I formulate also certain constitutive equations, which characterize the electromagnetic response of various media. The field equations coupled with the constitutive equations form a deterministic system of partial differential equations for the electromagnetic fields.

The set of electromagnetic constitutive equations for a vacuum is known as the system of Maxwell–Lorentz ether relations. An unsatisfactory feature of the classical theory is that the system of Maxwell–Lorentz ether relations is not invariant under a general Galilean transformation. As a result, the classical theory predicts that in a vacuum the speed of light (i.e., an electromagnetic wave) relative to the ether frame cannot be the same as that relative to a moving frame. This prediction is known to be inconsistent with experimental observations.

The above difficulty in the classical theory of electromagnetism is removed in the special theory of relativity, which uses the Minkowskian space–time as the model for the event world. In Chapter 6, I explain the structure of Minkowskian space–time in detail and reformulate the mathematical principles in electromagnetism to comply with the new model. Then I compare the results of the special relativistic theory with those of the classical theory.

Although the special theory of relativity removes the difficulty in the classical theory due to the transformation property of the Maxwell–Lorentz ether relations, it still depends on a preferred set of frames, known as the set of Lorentz frames. This set corresponds to the set of inertial frames in the classical theory. Both the Maxwell–Lorentz ether relations and Maxwell's field equations have the same forms relative to all Lorentz frames. Physically, the precise interpretation of the set of inertial frames or the set of Lorentz frames remains a major difficulty in the models.

In the classical theory it is understood that the departure of a frame from an inertial frame gives rise to an inertial acceleration field, which is indistinguishable from a gravitational field of the right type. For instance, if the frame is moving at a constant translational acceleration relative to an inertial frame, then its inertial acceleration field is a constant vector field and is indistinguishable from a constant gravitational field. Physically, there is associated with any frame of reference only an external acceleration field, which is the resultant of the inertial acceleration and the gravitational acceleration. An inertial frame is simply one in which the external acceleration is regarded as due to gravitation alone.

In the view of Einstein the external acceleration need not and cannot be decomposed intrinsically into a gravitational component and an inertial component; these components do not have separate physical significance. A relativistic model can be developed to account for the external acceleration directly. This model is given by the general theory of relativity.

In Chapter 7, I present the basic ideas of general relativity. This theory is based on the Minkowskian manifold as the model for the event world; a set of preferred frames is not required for this model. Unlike Minkowskian space–time, the Minkowskian manifold may have curvature associated with the Minkowskian metric. In accord with Einstein's interpretation, the Minkowskian metric plays the role of the gravitational potential and must satisfy a set of equations, known as Einstein's field equations. I summarize the results of these field equations for the problem of planetary orbits and the deflection of light and compare them with corresponding results in the classical theory.

I present also some interesting results in a recent theory by Toupin, who shows that a Minkowskian metric and an orientation on a 4-dimensional manifold give rise to two canonical relations among differential forms on the manifold. One of the two relations corresponds precisely to the system of Maxwell–Lorentz ether relations; I call the other relation the system of Nordström–Toupin ether relations. It turns out that the two canonical relations possess certain basic properties which may be used to determine the Minkowskian metric and the orientation.

In Chapter 8, I first reformulate Maxwell's equations in order to comply with the model of general relativity, and then discuss in detail Toupin's results on the connections between the system of Maxwell–Lorentz ether relations and the Minkowskian metric and the orientation.

Only a few topics in Part A, such as the balance principles and the field equations in continuum mechanics, are referred to in Part B for the purpose of comparing the results. Mathematical preliminaries for this Part are taken also mostly from the two-volume work *Introduction to Vectors and Tensors*,* published in this Series (Mathematical Concepts and Methods in Science and Engineering) in 1976.

I take this opportunity to thank Richard Toupin for providing me with an unpublished manuscript through the good office of Clifford Truesdell. That manuscript contains more details of his lectures given at Bressanone, Italy, in the summer of 1965. However, I have not referred directly to any result from that manuscript in this work. To my good friend, Jack Elliot, who also attended the Bressanone lectures, I wish to express my gratitude for his help in preparing some preprints of this and other works of mine.

This work is dedicated to my teacher, Clifford Truesdell, who has kept me honest by raising many critical questions and remarks, especially on the draft of Part B.

The support of the U.S. National Science Foundation in the preparation of this work is gratefully acknowledged.

C.-C. Wang

Houston, Texas

* R. M. Bowen and C.-C. Wang, Plenum Press, New York.

Contents of Part B
Electromagnetism and Gravitation

Contents of Part A
Analytical and Continuum Mechanics

5

Classical Theory of Electromagnetism

The classical theory of electromagnetism is formulated on the basis of a particular frame of reference, called the *rest frame* or the *ether frame*, in the Newtonian space–time. We develop this theory in four stages: first, electrostatics concerning the electric fields associated with a steady distribution of electric charges; second, magnetostatics concerning the magnetic fields associated with a steady distribution of electric currents; third, time-varying electromagnetism concerning the time-dependent electromagnetic fields associated with unsteady charge–current distributions; and fourth, electromagnetism for deforming media concerning the interaction of electromagnetic fields with mechanical stress and strain fields.

30. Classical Laws of Electrostatic Fields

In the classical theory of electromagnetism the event world is characterized by the Newtonian space–time $\mathscr{E}$ as in classical mechanics, but a particular frame of reference, called the *rest frame* or the *ether frame*, is assigned. It is generally agreed that this frame is an inertial frame as defined in classical mechanics. Relative to this frame $\mathscr{E}$ is represented in a definite way by the product space $\mathscr{S} \times \mathscr{R}$. All concepts in classical mechanics, such as forces and motions, are admitted directly into the mathematical formulation of electromagnetism.

We introduce first a primitive concept, called a *point charge*. Like a point mass or a particle in classical mechanics, a point charge may be characterized by a real number Q, called the *electric charge*, and by a point $\mathbf{y} \in \mathscr{S}$, called the *position* or the *location* of the point charge. In electrostatics we require that $\mathbf{y}$ be held fixed in $\mathscr{S}$ for all times t, i.e., the point charge does not move relative to the ether frame. Unlike the mass of a particle, the electric charge may be positive or negative. If the electric charges Q_1 and Q_2 have the same sign, then they are called *like charges*; otherwise, they are called *unlike charges* or *opposite charges.*

In the physical world the main phenomena associated with stationary electric charges are the *electrostatic fields*. Electric charges play two distinct roles with respect to electrostatic fields. First, the presence of any electric charge in space gives rise to an electrostatic field, and second, the presence of any electrostatic field (due to other electric charges in space) gives rise to a force acting on any electric charge as soon as that charge is put into the field. To balance the force due to the field, we may use a suitable support force. Indeed, the electric force acting on any point charge may be determined by the support force required to keep the charge in static equilibrium.

The quantitative concept of an electric field originated from Coulomb's empirical determination of the force acting on a point charge due to the presence of another point charge in space. Let Q_1 and Q_2 be located at the positions $\mathbf{y}_1$ and $\mathbf{y}_2$, respectively. Then the force $\mathbf{f}$ acting on Q_2 due to Q_1 is given by *Coulomb's law*:

$$\mathbf{f} = K \frac{Q_1 Q_2}{r^3} \mathbf{r}, \tag{30.1}$$

where $\mathbf{r} = \mathbf{y}_2 - \mathbf{y}_1$ denotes the position vector of $\mathbf{y}_2$ relative to $\mathbf{y}_1$, r denotes the length of $\mathbf{r}$, and K denotes a positive real number, whose value depends on the units used for electric charge and force. The fact that like charges repel and unlike charges attract each other is reflected by the positive sign of K. In the electrostatic units (esu) K is taken as unity. Then (30.1) reduces to

$$\mathbf{f} = \frac{Q_1 Q_2}{r^3} \mathbf{r} \qquad \text{(in esu)}. \tag{30.2}$$

Since the right-hand side of (30.2) is proportional to Q_2, we can define an *electric field* $\mathbf{E}$ in space due to the presence of a point charge Q by

$$\mathbf{E} = \mathbf{E}(\mathbf{x}) = \frac{Q}{r^3} \mathbf{r}, \tag{30.3}$$

where $\mathbf{r} = \mathbf{x} - \mathbf{y}$ denotes the position vector of $\mathbf{x}$ in the electric field relative to the position $\mathbf{y}$ of the point charge Q. Then from (30.2) $\mathbf{E}(\mathbf{x})$ is just the force acting on a unit charge if that charge is introduced at the position $\mathbf{x}$. It follows from (30.3) that $\mathbf{E}$ is a conservative vector field which may be expressed as the gradient of a potential, viz.,

$$\mathbf{E} = -\text{grad}\left(\frac{Q}{r}\right). \tag{30.4}$$

A point charge may be generalized to a *distribution of electric charges* in the same way that a particle is generalized to a body in mechanics. Thus the formulas (30.3) and (30.4) may be generalized to

$$\mathbf{E}(\mathbf{x}) = \int_{\mathscr{D}_q} \frac{q\mathbf{r}}{r^3}\, dv = -\text{grad}\left(\int_{\mathscr{D}_q} \frac{q}{r}\, dv\right), \tag{30.5}$$

where q denotes the charge density on some domain $\mathscr{D}_q$ in space, and where dv denotes the Euclidean element of volume at the position $\mathbf{y} = \mathbf{x} - \mathbf{r}$ in $\mathscr{D}_q$. As before we assume that a suitable field of support forces is present to keep the charges in static equilibrium.

The formula (30.5) shows that $\mathbf{E}$ is again a conservative field, and that the potential function φ is given by

$$\varphi = \varphi(\mathbf{x}) = \int_{\mathscr{D}_q} \frac{q}{r}\, dv. \tag{30.6}$$

A direct consequence of (30.5) is

$$\text{curl}\, \mathbf{E} = \mathbf{0}. \tag{30.7}$$

It should be noted that the formulas (30.5) and (30.6) are valid at all points $\mathbf{x}$ in space, including the points inside the domain $\mathscr{D}_q$. Indeed, if q is a bounded field, say q is continuous on the closure $\bar{\mathscr{D}}_q$, then q/r is an integrable field even though it has a singularity at the point $\mathbf{x} = \mathbf{y} \in \mathscr{D}_q$, where $r = 0$. This result is discussed in potential theory.[1] From that theory we have the general formula

$$\text{div}\left(\text{grad} \int_{\mathscr{D}_h} \frac{h}{r}\, dv\right) = -4\pi h(\mathbf{x}) \tag{30.8}$$

for any continuous function h, where dv has the meaning as before. Taking the divergence of (30.5) and using the general formula (30.8), we obtain the field equation

$$\text{div}\, \mathbf{E} = 4\pi q. \tag{30.9}$$

[1] See, for example, O. D. Kellogg, *Foundations of Potential Theory*, Dover edition, New York, 1953.

This result may be expressed in the integral form

$$\int_{\mathscr{D}} \operatorname{div} \mathbf{E}\, dx = \int_{\partial\mathscr{D}} \mathbf{E} \cdot \mathbf{n}\, d\sigma = 4\pi \int_{\mathscr{D}} q\, dx, \tag{30.10}$$

where $\mathscr{D}$ is any regular domain in space such that the divergence theorem holds. We call the surface integral, $\int_{\partial\mathscr{D}} \mathbf{E} \cdot \mathbf{n}\, d\sigma$, the *electric flux* through $\partial\mathscr{D}$. Then (30.10) implies that the electric flux is equal to 4π times the total charge enclosed by the surface $\partial\mathscr{D}$. This assertion is known as *Gauss' flux theorem.*

The mathematical model formulated so far has only limited applications for the following reason: In the physical world electric charges are present in two distinct forms: *free charges* and *bound charges*. Free charges are directly accessible to us in the sense that we can introduce or remove them freely at any position in space. Bound charges, on the other hand, are confined to the molecules of materials and are therefore not accessible to us. We can only observe the effects of the bound charges but we cannot introduce or remove them directly.

In some sense the distinction between the free charges and the bound charges is just like that between the body forces and the contact forces in a material body. We can introduce or remove body forces at will but we cannot do the same to the contact forces. Like the contact forces, which are due to the response of the material toward deformations, the bound charges are due to the response of the material toward electric fields. In order to apply Coulomb's formulas directly, we have to account for all electric charges—both the free charges and the bound charges. This situation is similar to that in continuum mechanics: in order to apply Cauchy's principles of motion, we have to account for all forces—both the body forces and the contact forces.

In continuum mechanics the difficulty of characterizing the contact forces is overcome by using the stress tensor, and the stress tensor is characterized mathematically by the constitutive equation. In classical electromagnetism we follow a similar procedure to characterize the bound charges.

We use the term *medium* to represent the physical notion of a material body or a vacuum. This term is chosen instead of the previous term *body manifold* in continuum mechanics mainly because in the case where the medium is a vacuum there are no identifiable body points. In electrostatics a medium is just a domain in space having a particular electrostatic response. We proceed now to formulate models for such response. First, the symbol q shall now refer to the free-charge density, and the symbol $\mathbf{E}$ shall refer to the electric field. We no longer require that $\mathbf{E}$ be determined by q through the formula (30.5), since $\mathbf{E}$ depends on both the free charges and the bound charges. At any point $\mathbf{x} \in \mathscr{S}$ the value $\mathbf{E}(\mathbf{x})$ shall have the direct physical meaning: the force acting on a unit point charge if that charge is introduced at the position $\mathbf{x}$. To characterize the bound charges we introduce a vector field $\mathbf{D}$ called the *electric displacement*. We require that $\mathbf{D}$ be related to q by

$$\operatorname{div} \mathbf{D} = 4\pi q \tag{30.11}$$

everywhere in space.

The physical meaning of the electric displacement has been explained by Faraday and by Maxwell on the basis of a molecular model for a dielectric material. We shall not go into the details of the explanation here, since we are primarily interested in the mathematical structure of the theory.

In some sense the electric displacement $\mathbf{D}$ is just like the stress tensor $\mathbf{T}$, and the field equation (30.11) is just like the equation of equilibrium $\operatorname{div} \mathbf{T} + \varrho\mathbf{b} = \mathbf{0}$. Mathematically, given the field q the equation (30.11) does not determine the field $\mathbf{D}$ uniquely, just like given the field $\varrho\mathbf{b}$ the equation of equilibrium does not determine the field $\mathbf{T}$ uniquely. To render the model deterministic, we introduce some boundary conditions and some constitutive equations as in continuum mechanics. We consider first the constitutive equations.

The constitutive equations in electrostatics are based on the assumption that at any point $\mathbf{x} \in \mathscr{S}$ the electric displacement $\mathbf{D}(\mathbf{x})$ is determined by the electric field $\mathbf{E}(\mathbf{x})$. Such a relation characterizes the electrostatic response of the medium at the point $\mathbf{x}$ and is independent of the field of free charge density q. This basic constitutive assumption has been explained by Faraday and by Maxwell on the basis of a molecular model mentioned before. For a vacuum we have $\mathbf{D} = \mathbf{E}$, since in this medium (30.11) reduces to (30.9), i.e., there are no bound charges in a vacuum. A medium is said to be (*electrically*) *isotropic* if $\mathbf{D}$ is proportional to $\mathbf{E}$,

$$\mathbf{D} = \varepsilon \mathbf{E}, \tag{30.12}$$

where ε is a positive material constant, known as the *permittivity*. A medium is called *anisotropic* if $\mathbf{D}$ is a linear function of $\mathbf{E}$,

$$\mathbf{D} = \boldsymbol{\epsilon} \mathbf{E}, \tag{30.13}$$

where $\boldsymbol{\epsilon}$ is a positive definite symmetric tensor, called the *permittivity tensor*. The relation (30.13) corresponds to three independent permittivities ε_i,

$i = 1, 2, 3$, associated with a principal basis $\{\mathbf{e}_i, i = 1, 2, 3\}$ of $\boldsymbol{\epsilon}$, i.e.,

$$D^i = \varepsilon_i E^i, \qquad i = 1, 2, 3, \tag{30.14}$$

where D^i and E^i denote the principal components of **D** and **E**. In particular, (30.13) reduces to (30.12) when $\boldsymbol{\epsilon}$ is completely degenerate and has only one distinct proper number ε.

Note. Like all constitutive equations, (30.12) and (30.13) are not laws of physics but are some convenient mathematical models which characterize the electrostatic response of some special classes of materials. For the most part classical electrostatics is formulated on the basis of (30.12) and (30.13). We can, of course, formulate mathematically some more general nonlinear constitutive equations as in the development of continuum mechanics. Some examples of constitutive equations more general than (30.13) are considered in the last section of this chapter.

In the physical interpretation the difference between **D** and **E**,

$$\mathbf{P} \equiv \mathbf{D} - \mathbf{E}, \tag{30.15}$$

may be attributed to the inaccessible bound charges in the medium. We call **P** the *polarization field*. In the absence of free charges the divergence of (30.15) implies that

$$\operatorname{div} \mathbf{E} = -\operatorname{div} \mathbf{P}. \tag{30.16}$$

This equation is comparable to the previous equation (30.9) for the electric field in a vacuum. Thus we may regard $-(1/4\pi) \operatorname{div} \mathbf{P}$ as a source for the electric field inside the medium.

Regardless of the nature of the bound charges **E** is always a conservative vector field, i.e.,

$$\operatorname{curl} \mathbf{E} = \mathbf{0}, \tag{30.17a}$$

$$\mathbf{E} = -\operatorname{grad} \varphi. \tag{30.17b}$$

However, because of the bound charges the potential function φ is generally not given by Coulomb's formula (30.6). Indeed, substituting (30.17b) into the constitutive equations (30.12) or (30.13), and then using the field equation (30.11), we get

$$\operatorname{div}(\boldsymbol{\epsilon} \operatorname{grad} \varphi) = -4\pi q. \tag{30.18}$$

Since $\boldsymbol{\epsilon}$ is positive definite and symmetric, this is an elliptic partial differential equation for φ assuming that the free charge density q is given. Suppose that $\boldsymbol{\epsilon}$ is smooth, that $\varphi \to 0$ at infinity, and that $q = 0$ outside a bounded domain $\mathscr{D}_q$. We can solve (30.18) and determine a unique solution φ. Then $\mathbf{E}$ is given by (30.17b), and $\mathbf{D}$ is given by (30.12) or by (30.13).

Note. For a vacuum $\boldsymbol{\epsilon} = \mathbf{I}$. Then the governing equation (30.18) reduces to the Poisson equation

$$\operatorname{div}(\operatorname{grad} \varphi) = \nabla^2 \varphi = -4\pi q. \tag{30.19}$$

In this case the solution φ is given by Coulomb's formula (30.6).

For different media the permittivity tensors are not the same. Hence the assumption that $\boldsymbol{\epsilon}$ be a smooth field in the whole space $\mathscr{S}$ is not realistic from the physical standpoint, since at the interface of two distinct media $\boldsymbol{\epsilon}$ is discontinuous. Then we must specify the boundary condition or the jump condition.

Let $\mathscr{U}$ be a surface in $\mathscr{S}$, and suppose that ψ is a field which is smooth on $\mathscr{S} - \mathscr{U}$ and has limits $\tilde{\psi}$ and $\hat{\psi}$ from the two sides of $\mathscr{U}$. Then we denote the difference $\tilde{\psi} - \hat{\psi}$ by $[\![\psi]\!]$, called the *jump* of ψ across the surface $\mathscr{U}$. Since $\mathbf{E}$ is the field of forces acting on a unit charge, (30.17b) implies that φ is the field of potential energy of a unit charge in the field $\mathbf{E}$, the potential energy at infinity being taken as zero. We assume that $[\![\varphi]\!] = 0$ across any surface $\mathscr{U}$. Then from the field equation (30.17b) we see that the tangential component of $\mathbf{E}$ must not suffer any jump discontinuity across the interface of two media, viz.,

$$[\![\mathbf{E}]\!] = [\![\mathbf{E} \cdot \mathbf{n}]\!]\mathbf{n}, \tag{30.20}$$

where $\mathbf{n}$ denotes the unit normal vector of the interface. This jump condition may be obtained also by using the field equation (30.17a) and Stokes' theorem on an elementary loop bridging the two sides of the interface.

Following the same idea, we use the divergence theorem on an elementary cylindrical domain bridging the two sides of the interface; then we obtain from (30.11)

$$[\![\mathbf{D}]\!] \cdot \mathbf{n} = [\![\mathbf{D} \cdot \mathbf{n}]\!] = 0. \tag{30.21}$$

Thus the normal component of $\mathbf{D}$ suffers no jump across the interface. In the derivation of (30.21) we have assumed that there is no free charge distribution on the interface. (This assumption is appropriate for the interface of dielectric materials.) From (30.20), (30.21), and (30.13) we can calculate the fields $\mathbf{D}$ and $\mathbf{E}$ on one side of the interface from the same fields on the other side.

Specifically, let $\tilde{\boldsymbol{\epsilon}}$, $\tilde{\mathbf{D}}$, $\tilde{\mathbf{E}}$, and $\hat{\boldsymbol{\epsilon}}$, $\hat{\mathbf{D}}$, $\hat{\mathbf{E}}$, denote the fields on the two sides of the interface. Then from (30.20)

$$\hat{\boldsymbol{\epsilon}}\tilde{\boldsymbol{\epsilon}}^{-1}\tilde{\mathbf{D}} - \hat{\mathbf{D}} = [\![\mathbf{E}\cdot\mathbf{n}]\!]\hat{\boldsymbol{\epsilon}}\mathbf{n}. \tag{30.22}$$

Taking the dot product of this equation with $\mathbf{n}$ and using (30.21), we get

$$\mathbf{n}\cdot\hat{\boldsymbol{\epsilon}}\tilde{\boldsymbol{\epsilon}}^{-1}\tilde{\mathbf{D}} - \mathbf{n}\cdot\tilde{\mathbf{D}} = [\![\mathbf{E}\cdot\mathbf{n}]\!]\mathbf{n}\cdot\hat{\boldsymbol{\epsilon}}\mathbf{n}. \tag{30.23}$$

Hence

$$[\![\mathbf{E}\cdot\mathbf{n}]\!] = \frac{\mathbf{n}\cdot(\hat{\boldsymbol{\epsilon}} - \tilde{\boldsymbol{\epsilon}})\tilde{\mathbf{E}}}{\mathbf{n}\cdot\hat{\boldsymbol{\epsilon}}\mathbf{n}}, \tag{30.24}$$

which implies that

$$\hat{\mathbf{E}} = \tilde{\mathbf{E}} - \left(\frac{\mathbf{n}\cdot(\hat{\boldsymbol{\epsilon}} - \tilde{\boldsymbol{\epsilon}})\tilde{\mathbf{E}}}{\mathbf{n}\cdot\hat{\boldsymbol{\epsilon}}\mathbf{n}}\right)\mathbf{n}. \tag{30.25}$$

Then $\hat{\mathbf{D}}$ is given by

$$\hat{\mathbf{D}} = \hat{\boldsymbol{\epsilon}}\tilde{\boldsymbol{\epsilon}}^{-1}\tilde{\mathbf{D}} - \left(\frac{\mathbf{n}\cdot(\hat{\boldsymbol{\epsilon}}\tilde{\boldsymbol{\epsilon}}^{-1} - \mathbf{I})\tilde{\mathbf{D}}}{\mathbf{n}\cdot\hat{\boldsymbol{\epsilon}}\mathbf{n}}\right)\hat{\boldsymbol{\epsilon}}\mathbf{n}. \tag{30.26}$$

Another kind of material medium considered in electrostatics is a *conductor*. This medium is defined by the condition that in its interior the fields $\mathbf{D}$ and $\mathbf{E}$ always vanish. Consequently, the potential function φ is constant throughout the medium, and there is no internal free charge distribution. At the interface of a conductor with a dielectric the jump condition (30.20) remains valid, except that we may remove the bracket notation and write simply

$$\mathbf{E} = (\mathbf{E}\cdot\mathbf{n})\mathbf{n} \tag{30.27}$$

for the field $\mathbf{E}$ in the dielectric medium, since the electric field vanishes in the conductor. The jump condition (30.21), however, is generally not valid, since there may be a free charge distribution on the interface. Indeed, by the argument leading toward (30.21), we now have

$$\mathbf{D}\cdot\mathbf{n} = 4\pi\tau, \tag{30.28}$$

where τ denotes the charge density per unit surface area.

Using the boundary conditions (30.25)–(30.28) together with the field equation (30.18), we can solve the potential function φ such that $\varphi \to 0$ at infinity. Then the electric field $\mathbf{E}$ may be determined by (30.17b), and the displacement field $\mathbf{D}$ may be determined by the constitutive equation (30.13) for each dielectric medium.

31. Steady Currents and Magnetic Induction

In the preceding section we considered the electric field **E** and the electric displacement field **D** associated with a stationary distribution of charges. We now consider a more general situation when charges are in motion. The existence of an electric current in a conducting wire due to a source of electric potential such as a voltaic battery is a known physical phenomenon. If the potential source is time independent, then it is known that the current in the conducting wire attached to the source is steady. Moreover, it is found from experiments that the current J is proportional to the electric potential V of the source,

$$J = \frac{1}{R} V, \tag{31.1}$$

where R is a constant, called the *resistance* of the conducting wire. The empirical formula (31.1) was discovered originally by Ohm and is now known as *Ohm's law*.

Suppose that there are steady currents J_1 and J_2 in a pair of wires in vacuum as shown in Fig. 3. It was observed in experiments by Ampère that there is a force acting on the second circuit due to the first circuit. Ampère obtained the formula

$$\mathbf{f} = KJ_1J_2 \oint\oint \frac{d\boldsymbol{\lambda}_2 \times (d\boldsymbol{\lambda}_1 \times \mathbf{r})}{r^3}, \tag{31.2}$$

where K is a constant depending on the units used. In the electromagnetic

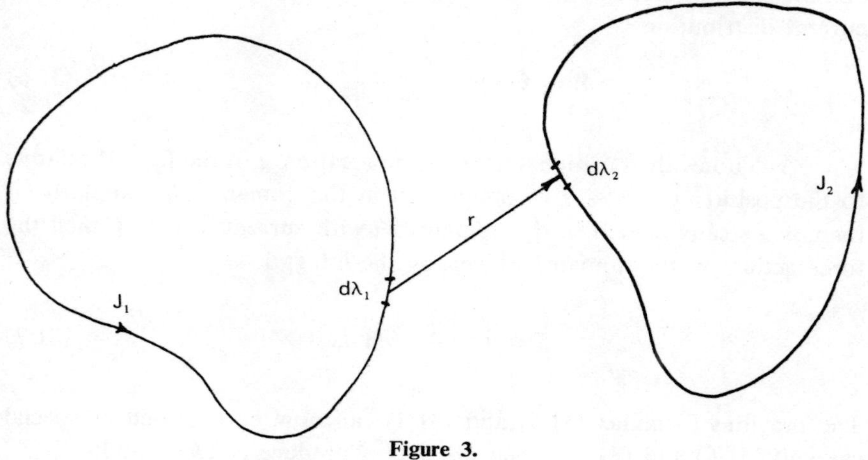

Figure 3.

units (emu) K is taken as unity. The unit of current thus obtained is called the *abampere*. This unit differs from the (esu) unit of charge per unit time by a constant c, whose dimension is that of the velocity. As we shall see in Section 34, c is just the speed of light (i.e., an electromagnetic wave) in vacuum and has an approximate value of 3×10^{10} cm/sec. Using the electrostatic units, $K = 1/c^2$, we have

$$\mathbf{f} = \frac{J_1 J_2}{c^2} \oint\oint \frac{d\boldsymbol{\lambda}_2 \times (d\boldsymbol{\lambda}_1 \times \mathbf{r})}{r^3} \qquad \text{(in esu)}, \tag{31.3}$$

which is known as *Ampère's law*.

From (31.3) we see that the force $\mathbf{f}$ may be regarded as the integral of a force field on the circuit 2:

$$\mathbf{f} = \frac{J_2}{c} \oint (d\boldsymbol{\lambda}_2 \times \mathbf{B}), \tag{31.4}$$

where

$$\mathbf{B} = \frac{J_1}{c} \oint \frac{d\boldsymbol{\lambda}_1 \times \mathbf{r}}{r^3}. \tag{31.5}$$

We call $\mathbf{B}$ the *magnetic induction* field, which is due to the steady current J_1 in the circuit 1.

The preceding analysis of the field induced by a steady current in a wire may be generalized to that of a distribution of steady currents in some domain $\mathscr{D}_{\mathbf{j}}$ in (vacuum) space. We denote the current density (in current per unit area) by $\mathbf{j}$. Then the magnetic induction $\mathbf{B}$ induced by the current distribution is

$$\mathbf{B} = \mathbf{B}(\mathbf{x}) = \int_{\mathscr{D}_{\mathbf{j}}} \frac{\mathbf{j} \times \mathbf{r}}{cr^3}\, dv, \tag{31.6}$$

where $\mathbf{r}$ denotes the position vector of the location $\mathbf{x}$ in the field $\mathbf{B}$ relative to the position $\mathbf{y} \equiv \mathbf{x} - \mathbf{r}$ of integration in the domain $\mathscr{D}_{\mathbf{j}}$. Similarly, if there is a steady field $\mathbf{B}$ in the domain $\mathscr{D}_{\bar{\mathbf{j}}}$ with current density $\bar{\mathbf{j}}$, then the force acting on the domain is given by the integral

$$\mathbf{f} = \int_{\mathscr{D}_{\bar{\mathbf{j}}}} \frac{\bar{\mathbf{j}}}{c} \times \mathbf{B}\, dx. \tag{31.7}$$

The previous formulas (31.5) and (31.4) can now be regarded as special cases of (31.6) and (31.7), when $\mathscr{D}_{\mathbf{j}}$ and $\mathscr{D}_{\bar{\mathbf{j}}}$ reduce to two circuits.

The formula (31.6) may be rewritten as

$$\mathbf{B} = -\int_{\mathscr{D}_\mathbf{j}} \frac{\mathbf{j}}{c} \times \operatorname{grad}\left(\frac{1}{r}\right) dv = \operatorname{curl} \int_{\mathscr{D}_\mathbf{j}} \frac{\mathbf{j}}{cr}\, dv, \tag{31.8}$$

which implies that $\mathbf{B}$ is a solenoidal vector field, viz.,

$$\operatorname{div} \mathbf{B} = 0, \tag{31.9a}$$

$$\mathbf{B} = \operatorname{curl} \mathbf{A}, \tag{31.9b}$$

where $\mathbf{A}$ is called a *vector potential* for $\mathbf{B}$. From (31.8) $\mathbf{A}$ is given by the integral

$$\mathbf{A} = \int_{\mathscr{D}_\mathbf{j}} \frac{\mathbf{j}}{cr}\, dv. \tag{31.10}$$

For representations of vector fields possessing special properties such as (31.9a) the reader is referred to Section 54, IVT-2. It is pointed out there that the vector potential of a solenoidal field is unique to within an additive conservative field only.

In any field of currents $\mathbf{j}$ electric charges are moving in space. We can visualize this situation as a flow of a distribution of point charges with velocity field $\mathbf{v}$. Then $\mathbf{j} = q\mathbf{v}$. Like the mass density ϱ of a continuum, the charge density q satisfies the continuity equation

$$\frac{\partial q}{\partial t} + \operatorname{div} \mathbf{j} = 0, \tag{31.11}$$

which is the field equation for the *law of conservation of charge*. The field equation (31.11) must be satisfied by any time-dependent charge and current fields in general. For steady fields the continuity equation reduces to

$$\operatorname{div} \mathbf{j} = 0. \tag{31.12}$$

We can use this field equation to show that the vector potential $\mathbf{A}$ given by (31.10) satisfies the additional condition

$$\operatorname{div} \mathbf{A} = 0, \tag{31.13}$$

which renders $\mathbf{A}$ unique provided that $\mathbf{A} \to \mathbf{0}$ at infinity.

To verify the condition (31.13), we take the divergence of **A** from (31.10), obtaining

$$\begin{aligned}\operatorname{div}\mathbf{A}(\mathbf{x}) &= \int_{\mathscr{D}_\mathbf{j}} \frac{\mathbf{j}(\mathbf{y})}{c}\cdot \operatorname{grad}_\mathbf{x}\left(\frac{1}{\|\mathbf{x}-\mathbf{y}\|}\right) dv \\ &= -\int_{\mathscr{D}_\mathbf{j}} \frac{\mathbf{j}(\mathbf{y})}{c}\cdot \operatorname{grad}_\mathbf{y}\left(\frac{1}{\|\mathbf{x}-\mathbf{y}\|}\right) dv \\ &= -\int_{\mathscr{D}_\mathbf{j}} \operatorname{div}_\mathbf{y}\left(\frac{\mathbf{j}(\mathbf{y})}{c\|\mathbf{x}-\mathbf{y}\|}\right) dv = -\int_{\partial\mathscr{D}_\mathbf{j}} \frac{\mathbf{j}(\mathbf{y})}{c\|\mathbf{x}-\mathbf{y}\|}\cdot \mathbf{n}(\mathbf{y})\, d\sigma_\mathbf{y},\end{aligned} \tag{31.14}$$

where we have used the divergence theorem and the continuity equation (31.12). Now the surface integral on the right-hand side of (31.14) vanishes, since by assumption the current field **j** is confined within a bounded domain.

Note. The preceding argument is valid for a smooth steady field of currents only, since in the derivation of (31.14) we have used the divergence theorem and the continuity equation (31.12). A more general condition for an unsteady field is considered in Section 32.

Taking the curl of the field **B** from (31.8), we get

$$\begin{aligned}\operatorname{curl}\mathbf{B} &= \operatorname{curl}\operatorname{curl}\int_{\mathscr{D}_\mathbf{j}} \frac{\mathbf{j}}{cr}\, dv \\ &= -\int_{\mathscr{D}_\mathbf{j}} \frac{\mathbf{j}}{c}\nabla^2\left(\frac{1}{r}\right) dv + \operatorname{grad}\operatorname{div}\mathbf{A} = 4\pi\left(\frac{\mathbf{j}}{c}\right),\end{aligned} \tag{31.15}$$

where we have used the general formula (30.8). Clearly (31.15) is consistent with (31.12). The field equations (31.9a) and (31.15) determine uniquely the field **B** if $\mathbf{B}\to\mathbf{0}$ at infinity. The solution is given by Ampère's formula (31.6). By using Stokes' theorem, we can rewrite the field equation (31.15) in the integral form

$$\oint_{\boldsymbol{\lambda}} \mathbf{B}\cdot d\boldsymbol{\lambda} = 4\pi\frac{J}{c}, \tag{31.16}$$

where J is the total current encircled by the circuit $\boldsymbol{\lambda}$, i.e.,

$$J = \int_{\mathscr{U}} \mathbf{j}\cdot\mathbf{n}\, d\sigma, \tag{31.17}$$

where **n** is the positive unit normal of the surface $\mathscr{U}$ such that $\partial\mathscr{U}$ is the circuit $\boldsymbol{\lambda}$ on the left-hand side of (31.16). The integral form (31.16) is known as *Ampère's circuit law*.

So far we have formulated the magnetic induction field **B** associated with some steady currents in (vacuum) space. This formulation has only limited applications for the same reason as before: In the physical world currents are present in two distinct forms: *free currents* and *bound currents* (or *magnetization currents*). Free currents are accessible to us, but bound currents are confined to the molecules of materials and are therefore not accessible to us. In order to apply Ampère's formulas directly, we must account for all currents in space—both the free currents and the bound currents. This difficulty may be overcome in the following way as before:

First, the symbol **j** shall now refer to the free-current density, and the symbol **B** shall refer to the magnetic induction field. We no longer require that **B** be determined by **j** through Ampère's formula (31.6), since **B** depends on both the free currents and the bound currents. The physical meaning of $\mathbf{B}(\mathbf{x})$ at any point $\mathbf{x} \in \mathscr{S}$ is given directly by the force **f** acting on any current according to the formula (31.7), when that current is introduced into the field at the position **x**. To characterize the bound currents in the material medium, we introduce a vector field **H**, called the *magnetic field*. We require that the field **H** be related to the field **j** by

$$\operatorname{curl} \mathbf{H} = 4\pi \frac{\mathbf{j}}{c} \tag{31.18}$$

The term *medium* is used here in a similar sense as before: a domain in $\mathscr{S}$ having a particular magnetostatic response. When the medium is a vacuum, we have $\mathbf{H} = \mathbf{B}$, and (31.18) reduces to (31.15), since there is no bound current in a vacuum. For a medium of magnetic material we assume that $\mathbf{H}(\mathbf{x})$ is determined by $\mathbf{B}(\mathbf{x})$. This constitutive relation characterizes mathematically the magnetic response of the medium and is independent of the free-current density **j**. A medium is said to be (*magnetically*) *isotropic* if **H** is proportional to **B**:

$$\mathbf{H} = \frac{1}{\mu} \mathbf{B}, \qquad \mathbf{B} = \mu \mathbf{H}, \tag{31.19}$$

where μ is a positive material constant, called the *permeability*. A medium is called anisotropic if **H** is given by a linear transformation of **B**:

$$\mathbf{H} = \boldsymbol{\mu}^{-1}\mathbf{B}, \qquad \mathbf{B} = \boldsymbol{\mu}\mathbf{H}, \tag{31.20}$$

where $\boldsymbol{\mu}$ is a positive definite symmetric tensor, called the *permeability tensor*. As before this tensor corresponds to three independent permeabilities μ_i, $i = 1, 2, 3$, associated with a principal basis of $\boldsymbol{\mu}$.

Note. As remarked before, the constitutive relations (31.19) or (31.20) are not laws of physics but are just some convenient mathematical models which characterize the magnetic response of some materials. Ferromagnetic materials such as iron and nickel may have permanent magnetization currents. For such materials the constitutive relations (31.19) or (31.20) are not applicable. In fact ferromagnetic materials often exhibit a memory effect known as *magnetic hysteresis*. Then **H** is not even given by a single-valued function of **B**. In this section we shall limit our attention to constitutive relations of the forms (31.19) or (31.20) only.

The difference between **B** and **H**,

$$\mathbf{M} = \mathbf{B} - \mathbf{H}, \tag{31.21}$$

is called the *magnetization*. In the physical interpretation this concept may be attributed to the inaccessible bound currents in the medium. For an isotropic medium we write also

$$\mathbf{M} = (\mu - 1)\mathbf{H} = \chi\mathbf{H} = (1 - 1/\mu)\mathbf{B}, \tag{31.22}$$

where χ is called the *magnetic susceptibility*. If $\chi > 0$, we call the medium *paramagnetic*; otherwise, we call the medium *diamagnetic*.

Regardless of the nature of the magnetization currents in the medium the field **B** is always a solenoidal field. However, the vector potential **A** need not be given by the formula (31.10), since **B** depends on both the free currents as well as the inaccessible magnetization currents. In the absence of free currents the magnetic field is a conservative field:

$$\operatorname{curl} \mathbf{H} = \mathbf{0}. \tag{31.23}$$

Also, from (31.21) and (31.9)

$$\operatorname{div} \mathbf{H} = -\operatorname{div} \mathbf{M}. \tag{31.24}$$

These two equations are comparable to the equations (30.7) and (30.9). Thus $-(1/4\pi)\operatorname{div} \mathbf{M}$ may be regarded as a source for the field **H** inside the medium. This source is attributed to the inaccessible magnetization current. The condition that **B** is always solenoidal means that there is no free magnetic source in space.

From (31.18) the magnetic field **H** may be determined to within an additive conservative field by the free-current distribution **j**. We set

$$\mathbf{H} = \mathbf{H}_0 - \operatorname{grad} \psi, \tag{31.25}$$

where $\mathbf{H}_0$ is fixed and satisfies the field equation

$$\operatorname{div} \mathbf{H}_0 = 4\pi \frac{\mathbf{j}}{c}, \tag{31.26}$$

and where ψ is an unknown function. To determine the function ψ, we use the constitutive equation (31.20) and the field equation (31.9a), obtaining

$$\operatorname{div}(\boldsymbol{\mu}(\mathbf{H}_0 - \operatorname{grad} \psi)) = 0. \tag{31.27}$$

Hence

$$\operatorname{div}(\boldsymbol{\mu} \operatorname{grad} \psi) = \operatorname{div}(\boldsymbol{\mu}\mathbf{H}_0), \tag{31.28}$$

which has the same form as (30.18) and is an elliptic partial differential equation for the unknown function ψ.

As before **B** and **H** must satisfy certain jump conditions at the interface of two distinct magnetic media. The results are

$$[\![\mathbf{B}]\!] \cdot \mathbf{n} = [\![\mathbf{B} \cdot \mathbf{n}]\!] = 0, \tag{31.29}$$

and

$$[\![\mathbf{H}]\!] = [\![\mathbf{H} \cdot \mathbf{n}]\!]\mathbf{n}, \tag{31.30}$$

provided that there are no free-current distributions on the interface. From these jump conditions we can calculate the fields $\hat{\mathbf{B}}$ and $\hat{\mathbf{H}}$ on one side of the interface from the fields $\tilde{\mathbf{B}}$ and $\tilde{\mathbf{H}}$ on the other side, viz.,

$$\hat{\mathbf{H}} = \tilde{\mathbf{H}} - \left(\frac{\mathbf{n} \cdot (\hat{\boldsymbol{\mu}} - \tilde{\boldsymbol{\mu}})\tilde{\mathbf{H}}}{\mathbf{n} \cdot \hat{\boldsymbol{\mu}}\mathbf{n}}\right)\mathbf{n}, \tag{31.31}$$

and

$$\hat{\mathbf{B}} = \hat{\boldsymbol{\mu}}\tilde{\boldsymbol{\mu}}^{-1}\tilde{\mathbf{B}} - \left(\frac{\mathbf{n} \cdot (\hat{\boldsymbol{\mu}}\tilde{\boldsymbol{\mu}}^{-1} - \mathbf{I})\tilde{\mathbf{B}}}{\mathbf{n} \cdot \hat{\boldsymbol{\mu}}\mathbf{n}}\right)\hat{\boldsymbol{\mu}}\mathbf{n}. \tag{31.32}$$

As usual we assume that $\psi \to 0$ at infinity. Then ψ is determined by the field equation (31.28) and the boundary conditions (31.29)–(31.32). From the solution ψ the magnetic field **H** may be determined by (31.25) and the magnetic induction **B** may be determined by the constitutive equation (31.20) for each magnetic medium.

32. Time-Dependent Electromagnetic Fields, Maxwell's Equations

In the preceding two sections we have formulated the laws governing time-independent electric fields and magnetic fields. Now we develop the more general theory of unsteady fields.

We recall first that for a time-dependent charge field q and a time-dependent current field $\mathbf{j}$ the conservation of charge requires the continuity equation

$$\frac{\partial q}{\partial t} + \operatorname{div} \mathbf{j} = 0. \tag{32.1}$$

Clearly this equation is not consistent with the field equation (31.18) unless $\partial q/\partial t$ vanishes. Hence we cannot admit (31.18) directly into the theory of unsteady fields. This difficulty may be resolved by using Maxwell's concept of *displacement current* in the following way:

First, we admit the field equation (30.11) for unsteady fields, i.e., we require that the instantaneous values of q and $\mathbf{D}$ be related by (30.11) at each time t. Using that field equation, we can rewrite the continuity equation (32.1) as

$$\operatorname{div}\left(\mathbf{j} + \frac{1}{4\pi} \frac{\partial \mathbf{D}}{\partial t}\right) = 0, \tag{32.2}$$

where the term $(1/4\pi)\partial\mathbf{D}/\partial t$ is called the *displacement current*. The physical meaning of this term has been explained by Maxwell.

Next, Maxwell generalized Ampère's law (30.11) to

$$\operatorname{curl} \mathbf{H} = 4\pi \frac{\mathbf{j}}{c} + \frac{1}{c} \frac{\partial \mathbf{D}}{\partial t} \tag{32.3}$$

for the unsteady case in general. The field equation (32.3) has been verified by experiments and is consistent with the continuity equation (32.1).

In Section 31 we discussed the magnetic induction due to a steady current. We pointed out that there is a force acting on a current if that current is located in a magnetic induction field. Thus the relation of the current to the magnetic induction is just like that of the charge to the electric field. Now we consider still another kind of electromagnetic induction, which was observed from experiments by Faraday.

Faraday discovered that an electric field is induced around a circuit when the magnetic induction field $\mathbf{B}$ through the surface bounded by the circuit is changing with respect to time. Faraday's law of induction asserts that the line integral of the electric field around any circuit $\boldsymbol{\lambda}$ is given by

$$\oint_{\boldsymbol{\lambda}} \mathbf{E} \cdot d\boldsymbol{\lambda} = -\frac{1}{c} \frac{d}{dt} \int_{\mathscr{U}} \mathbf{B} \cdot \mathbf{n} \, d\sigma, \tag{32.4}$$

where $\mathbf{n}$ denotes the positive unit normal of the surface $\mathscr{U}$ such that $\partial\mathscr{U}$ is the circuit $\boldsymbol{\lambda}$.

In Faraday's original experiments a conducting wire was used to form the circuit, and the electric field was observed from the resulting current in the wire. Maxwell pointed out that the electric field was induced in the whole space not just in the wire. Hence (32.4) is valid for all loops $\boldsymbol{\lambda}$. By using Stokes' theorem, we then have

$$\int_{\mathscr{U}} \operatorname{curl} \mathbf{E} \cdot \mathbf{n} \, d\sigma = -\int_{\mathscr{U}} \frac{1}{c} \frac{\partial \mathbf{B}}{\partial t} \cdot \mathbf{n} \, d\sigma \tag{32.5}$$

for any oriented surface $\mathscr{U}$. Assuming that the integrands in (32.5) are continuous, we obtain the field equation for the electric field $\mathbf{E}$ in the unsteady case

$$\operatorname{curl} \mathbf{E} = -\frac{1}{c} \frac{\partial \mathbf{B}}{\partial t}, \tag{32.6}$$

which shows that $\mathbf{E}$ is generally not a conservative field.

The four field equations

$$\operatorname{div} \mathbf{D} = 4\pi q, \tag{32.7a}$$

$$\operatorname{curl} \mathbf{E} = -\frac{1}{c} \frac{\partial \mathbf{B}}{\partial t}, \tag{32.7b}$$

$$\operatorname{div} \mathbf{B} = 0, \tag{32.7c}$$

$$\operatorname{curl} \mathbf{H} = \frac{4\pi}{c} \mathbf{j} + \frac{1}{c} \frac{\partial \mathbf{D}}{\partial t}, \tag{32.7d}$$

valid for time-dependent charge, current, electric, and magnetic fields in general, are known as *Maxwell's equations*.

From the definition of the electric field the force acting on a point charge Q in an electric field $\mathbf{E}$ is given by $Q\mathbf{E}$. From Ampère's law if the point charge Q is moving with a velocity $\mathbf{v}$ in a magnetic induction field $\mathbf{B}$, then the force acting on the current $Q\mathbf{v}$ is $(Q\mathbf{v}/c) \times \mathbf{B}$. The total force acting on the moving point charge is then given by

$$\mathbf{f} = Q\left(\mathbf{E} + \frac{\mathbf{v}}{c} \times \mathbf{B}\right). \tag{32.8}$$

If there is a distribution of charges q moving with a velocity field $\mathbf{v}$, then (32.8) may be generalized to a field equation

$$\mathbf{f} = q\left(\mathbf{E} + \frac{\mathbf{v}}{c} \times \mathbf{B}\right) = q\mathbf{E} + \frac{\mathbf{j}}{c} \times \mathbf{B}, \tag{32.9}$$

where $\mathbf{f}$ denotes the force density. This formula was observed originally by Lorentz, so $\mathbf{f}$ is known as the *Lorentz force*. Maxwell's equations (32.7) together with Lorentz's formula (32.9) form the basis for the mathematical analysis of electromagnetic phenomena in the unsteady case.

As in the steady case considered in the preceding two sections, the field equations (32.7) may be solved by using some potential functions. First, (32.7c) implies as before the existence of a vector potential $\mathbf{A}$ such that

$$\mathbf{B} = \operatorname{curl} \mathbf{A}, \tag{32.10}$$

where $\mathbf{A}$ is unique to within an additive time-dependent conservative field. Substituting (32.10) into (32.7b), we get

$$\operatorname{curl}\left(\mathbf{E} + \frac{1}{c}\frac{\partial \mathbf{A}}{\partial t}\right) = \mathbf{0}, \tag{32.11}$$

which implies that

$$\mathbf{E} + \frac{1}{c}\frac{\partial \mathbf{A}}{\partial t} = -\operatorname{grad} \varphi, \tag{32.12}$$

where φ is unique to within an additive function of time.

Now suppose that the medium satisfies the linear constitutive relations (30.13) and (31.20) in the unsteady case also. Then in terms of the potentials φ and $\mathbf{A}$ we have

$$\mathbf{D} = -\boldsymbol{\epsilon}\left(\operatorname{grad} \varphi + \frac{1}{c}\frac{\partial \mathbf{A}}{\partial t}\right), \qquad \mathbf{H} = \boldsymbol{\mu}^{-1} \operatorname{curl} \mathbf{A}. \tag{32.13}$$

Substituting these representations into the remaining field equations (32.7a) and (32.7d), we get

$$\begin{gathered}\operatorname{div}(\boldsymbol{\epsilon} \operatorname{grad} \varphi) + \frac{1}{c}\operatorname{div}\left(\boldsymbol{\epsilon}\frac{\partial \mathbf{A}}{\partial t}\right) = -4\pi q, \\ \frac{1}{c}\frac{\partial}{\partial t}(\boldsymbol{\epsilon} \operatorname{grad} \varphi) + \frac{1}{c^2}\frac{\partial}{\partial t}\left(\boldsymbol{\epsilon}\frac{\partial \mathbf{A}}{\partial t}\right) + \operatorname{curl}(\boldsymbol{\mu}^{-1} \operatorname{curl} \mathbf{A}) = \frac{4\pi}{c}\mathbf{j}.\end{gathered} \tag{32.14}$$

These are the governing equations for the potentials φ and $\mathbf{A}$ for any charge and current distributions q and $\mathbf{j}$ [which must satisfy the conservation law (32.1); otherwise, (32.14a) is not consistent with (32.14b)].

For simplicity we consider a homogeneous isotropic medium. Then $\boldsymbol{\epsilon}$ and $\boldsymbol{\mu}$ reduce to the constants $\varepsilon\mathbf{I}$ and $\mu\mathbf{I}$. In this special case (32.14) becomes

$$\nabla^2\varphi + \frac{1}{c}\operatorname{div}\frac{\partial\mathbf{A}}{\partial t} = -\frac{4\pi}{\varepsilon}q, \tag{32.15a}$$

$$\frac{\varepsilon\mu}{c}\operatorname{grad}\frac{\partial\varphi}{\partial t} + \frac{\varepsilon\mu}{c^2}\frac{\partial^2\mathbf{A}}{\partial t^2} + \operatorname{curl}\operatorname{curl}\mathbf{A} = \frac{4\pi\mu}{c}\mathbf{j}. \tag{32.15b}$$

Now we claim that, without loss of generality, we can impose on the potentials φ and $\mathbf{A}$ the following *Lorentz condition*:

$$\operatorname{div}\mathbf{A} + \frac{\mu\varepsilon}{c}\frac{\partial\varphi}{\partial t} = 0. \tag{32.16}$$

Indeed, if $\mathbf{A}$ and φ fail to satisfy the preceding condition, then we can replace them by

$$\bar{\mathbf{A}} = \mathbf{A} + \operatorname{grad}\psi, \qquad \bar{\varphi} = \varphi - \frac{1}{c}\frac{\partial\psi}{\partial t}, \tag{32.17}$$

which still satisfy the representations (32.10) and (32.12). We may choose ψ in such a way that the new potentials $\bar{\mathbf{A}}$ and $\bar{\varphi}$ satisfy the condition (32.16). Specifically, $\bar{\mathbf{A}}$ and $\bar{\varphi}$ obey the Lorentz condition if and only if ψ satisfies the partial differential equation

$$\nabla^2\psi - \frac{\mu\varepsilon}{c^2}\frac{\partial^2\psi}{\partial t^2} = -\operatorname{div}\mathbf{A} - \frac{\mu\varepsilon}{c}\frac{\partial\varphi}{\partial t}, \tag{32.18}$$

which may be solved by the retarded potential:

$$\psi(\mathbf{x}, t) = \int_{\mathscr{D}_\lambda}\frac{\lambda(\mathbf{y}, t - \xi r)}{r}\,dv, \tag{32.19}$$

where

$$r = \|\mathbf{x} - \mathbf{y}\|, \qquad \xi = \frac{(\mu\varepsilon)^{1/2}}{c}, \qquad \lambda = \frac{1}{4\pi}\left(\operatorname{div}\mathbf{A} + \frac{\mu\varepsilon}{c}\frac{\partial\varphi}{\partial t}\right). \tag{32.20}$$

To prove that (32.19) gives a solution to the equation

$$\nabla^2\psi - \xi^2\frac{\partial^2\psi}{\partial t^2} = -4\pi\lambda, \tag{32.21}$$

we note that for any function ζ of $r = \| \mathbf{x} - \mathbf{y} \|$ the Laplacian is given by

$$\nabla^2 \zeta(r) = \frac{1}{r} \frac{d^2}{dr^2} (r\zeta(r)). \tag{32.22}$$

Also, for any function η of $t - \xi r$

$$\frac{\partial^2}{\partial r^2} \eta(t - \xi r) = \xi^2 \frac{\partial^2}{\partial t^2} \eta(t - \xi r). \tag{32.23}$$

Differentiating (32.19) under the integral and using (30.18), (32.22), and (32.23), we obtain

$$\begin{aligned} \nabla^2 \psi &= \int_{\mathscr{D}_\lambda} \lambda \nabla^2 \left(\frac{1}{r}\right) dv + \xi^2 \frac{\partial^2}{\partial t^2} \int_{\mathscr{D}_\lambda} \frac{\lambda(\mathbf{y}, t - \xi r)}{r} dv \\ &= -4\pi\lambda + \xi^2 \frac{\partial^2 \psi}{\partial t^2} = -4\pi\lambda + \frac{\mu\varepsilon}{c^2} \frac{\partial^2 \psi}{\partial t^2}. \end{aligned} \tag{32.24}$$

Thus the validity of the solution (32.19) is proved.

Now assuming that φ and $\mathbf{A}$ obey the Lorentz condition (32.16), we can rewrite (32.15a) as

$$\nabla^2 \varphi - \frac{\mu\varepsilon}{c^2} \frac{\partial^2 \varphi}{\partial t^2} = -\frac{4\pi}{\varepsilon} q. \tag{32.25}$$

Using the vector identity

$$\text{curl curl } \mathbf{A} = \text{grad}(\text{div } \mathbf{A}) - \nabla^2 \mathbf{A}, \tag{32.26}$$

we can rewrite (32.15b) similarly as

$$\nabla^2 \mathbf{A} - \frac{\mu\varepsilon}{c^2} \frac{\partial^2 \mathbf{A}}{\partial t^2} = -\frac{4\pi\mu}{c} \mathbf{j}. \tag{32.27}$$

Notice that the system (32.15) is now decoupled into two separate equations (32.25) and (32.27) for φ and $\mathbf{A}$, respectively.

Since (32.25) and (32.27) have the same form as (32.21), they can be solved by the retarded potentials

$$\varphi(\mathbf{x}, t) = \int_{\mathscr{D}_q} \frac{q(\mathbf{y}, t - \xi r)}{\varepsilon r} dv \tag{32.28}$$

and

$$\mathbf{A}(\mathbf{x}, t) = \int_{\mathscr{D}_\mathbf{j}} \frac{\mu \mathbf{j}(\mathbf{y}, t - \xi r)}{c r} dv, \tag{32.29}$$

which satisfy the Lorentz condition. From these solutions we can then determine the fields **B** and **E** by the representations (32.10) and (32.12).

At the interface of two distinct media the instantaneous fields **E**, **D**, **B**, and **H** are not continuous. We assume that the right-hand side of Maxwell's equations are integrable fields with respect to the Euclidean volume. Then by using (32.7b) and (32.7d) and Stokes' theorem on an elementary loop bridging the two sides of the interface, we see that the tangential components of **E** and **H** suffer no jump, i.e., we still have the jump conditions (30.20) and (31.30) in the unsteady case. Likewise by using (32.7a) and (32.7c) and the divergence theorem on an elementary cylindrical domain bridging the two sides of the interface, we see that the normal components of **D** and **B** suffer no jump, so that (30.21) and (31.29) still hold. Then following the procedure as before, we can express the values $\hat{\mathbf{E}}$, $\hat{\mathbf{D}}$, $\hat{\mathbf{B}}$, and $\hat{\mathbf{H}}$ on one side of the interface in terms of the values $\tilde{\mathbf{E}}$, $\tilde{\mathbf{D}}$, $\tilde{\mathbf{B}}$, and $\tilde{\mathbf{H}}$ on the other side, i.e., (30.25), (30.26), (31.31), and (31.32) still hold at each instant t.

So far we have developed the classical theory of electromagnetism in three stages: first electrostatics, then magnetostatics with steady currents, and finally unsteady electromagnetism. In order to distinguish a steady field from an unsteady field, we must use a particular frame of reference on the Newtonian space–time $\mathscr{E}$. This frame is known as the *rest frame* or the *ether frame* of classical electromagnetism. The requirement that such a frame must exist is a basic difficulty in the classical theory.

To illustrate this point, we consider a change of frame by a uniform translation with velocity **u**, i.e., the position vector $\mathbf{x}'$ of an event relative to the moving frame is related to the position vector **x** of the same event relative to the rest frame by

$$\mathbf{x}' = \mathbf{x} - \mathbf{u}t. \tag{32.30}$$

For simplicity we assume that the two frames use the same Newtonian time t. From (32.30) the velocity field $\mathbf{v}'$ relative to the moving frame is related to the velocity field **v** relative to the rest frame by

$$\mathbf{v}' = \mathbf{v} - \mathbf{u}. \tag{32.31}$$

Now conceptually it is very natural to regard the charge density q, like the mass density ϱ in Newtonian mechanics, as being a frame-indifferent quantity, viz.,

$$q' = q. \tag{32.32}$$

Then the current density $\mathbf{j}'$ relative to the moving frame is related to that in the rest frame by

$$\mathbf{j}' = \mathbf{j} - q\mathbf{u}. \tag{32.33}$$

Next, since the field $\mathbf{D}$ and the field $\mathbf{B}$ are used to characterize the free charge distribution and the free magnetic source, which are both assumed to be frame-indifferent, we have

$$\mathbf{D}' = \mathbf{D}, \tag{32.34a}$$

$$\mathbf{B}' = \mathbf{B}. \tag{32.34b}$$

Then the partial time derivatives $\partial \mathbf{D}'/\partial t$ and $\partial \mathbf{B}'/\partial t$ are related to the partial time derivatives $\partial \mathbf{D}/\partial t$ and $\partial \mathbf{B}/\partial t$ by

$$\frac{\partial \mathbf{D}'}{\partial t} = \frac{\partial \mathbf{D}}{\partial t} + (\text{grad } \mathbf{D})\mathbf{u}, \qquad \frac{\partial \mathbf{B}'}{\partial t} = \frac{\partial \mathbf{B}}{\partial t} + (\text{grad } \mathbf{B})\mathbf{u}, \tag{32.35}$$

where the extra terms on the right-hand side are due to the fact that in calculating the time derivatives, $\mathbf{x}'$ is held fixed in the moving frame while $\mathbf{x}$ is held fixed in the rest frame. That is, (32.34a) means explicitly that

$$\mathbf{D}'(\mathbf{x}', t) = \mathbf{D}(\mathbf{x}, t) = \mathbf{D}(\mathbf{x}' + \mathbf{u}t, t). \tag{32.36}$$

It follows from the transformation laws (32.30)–(32.36) that Maxwell's equations and Lorentz's formula become

$$\begin{aligned} \text{div}' \, \mathbf{D}' &= 4\pi q', \qquad \text{curl}'\left(\mathbf{E} + \frac{\mathbf{u}}{c} \times \mathbf{B}\right) = -\frac{1}{c}\frac{\partial \mathbf{B}'}{\partial t}, \\ \text{div}' \, \mathbf{B}' &= 0, \qquad \text{curl}'\left(\mathbf{H} - \frac{\mathbf{u}}{c} \times \mathbf{B}\right) = \frac{4\pi}{c}\mathbf{j}' + \frac{1}{c}\frac{\partial \mathbf{D}'}{\partial t}, \end{aligned} \tag{32.37}$$

and

$$\mathbf{f} = q'\left(\mathbf{E} + \frac{\mathbf{u}}{c} \times \mathbf{B} + \frac{\mathbf{v}'}{c} \times \mathbf{B}'\right), \tag{32.38}$$

where the prime on div′ and curl′ refers the operators to the moving frame. Comparing (32.37) and (32.38) with (32.7) and (32.9), we see that the forms of Maxwell's equations and Lorentz's formula are preserved by the change of frame if we define the electric field $\mathbf{E}'$ and the magnetic field $\mathbf{H}'$ in the moving frame by

$$\mathbf{E}' = \mathbf{E} + \frac{\mathbf{u}}{c} \times \mathbf{B}, \qquad \mathbf{H}' = \mathbf{H} - \frac{\mathbf{u}}{c} \times \mathbf{D}. \tag{32.39}$$

These transformation laws, however, will not preserve the constitutive relations, which should be frame-indifferent.

In fact while $\mathbf{D} = \mathbf{E}$ and $\mathbf{B} = \mathbf{H}$ for a vacuum in the rest frame, according to (32.34) and (32.39) we have $\mathbf{D}' \neq \mathbf{E}'$ and $\mathbf{B}' \neq \mathbf{H}'$ for the same vacuum in a moving frame. Thus the constitutive relations do not just refer to the response of a medium but depend also on the motion of the frame of reference relative to the ether frame.

Because of this difficulty a better mathematical model for electromagnetism has been formulated in the context of the theory of relativity. We shall discuss the relativistic model in the next three chapters.

The transformation laws (32.39) suggest that we should consider more general linear constitutive relations of the symmetric form

$$\mathbf{E} = \boldsymbol{\varkappa}\mathbf{D} + \boldsymbol{\chi}\mathbf{B}, \qquad \mathbf{H} = \boldsymbol{\chi}^T\mathbf{D} + \boldsymbol{\nu}\mathbf{B}, \tag{32.40}$$

or the inverse form

$$\mathbf{D} = \boldsymbol{\epsilon}\mathbf{E} + \boldsymbol{\xi}\mathbf{H}, \qquad \mathbf{B} = \boldsymbol{\xi}^T\mathbf{E} + \boldsymbol{\mu}\mathbf{H}. \tag{32.41}$$

A medium defined by such a system of constitutive relations may be called *bianisotropic*, where the prefix "bi" refers to the cross coupling of the electric and the magnetic fields. [The symmetry of the coefficient matrices in (32.40) and (32.41) is known as the *lossless condition.*] When $\boldsymbol{\epsilon}$, $\boldsymbol{\xi}$, and $\boldsymbol{\mu}$ reduce to scalars, viz.,

$$\mathbf{D} = \varepsilon\mathbf{E} + \xi\mathbf{H}, \qquad \mathbf{B} = \xi\mathbf{E} + \mu\mathbf{H}, \tag{32.42}$$

the medium is called *biisotropic*.

Historically biisotropic or bianisotropic media were considered in the theory of electromagnetism for moving media; cf. Section 35. It has been shown recently, however, that the cross coupling of electromagnetic fields may exist in certain types of magnetic crystals. The first experimental observations of such cross coupling were made in 1960 in the antiferromagnetic chromium oxide. The system of constitutive relations for this medium is of the form

$$\begin{aligned}
\begin{pmatrix} D^1 \\ D^2 \\ D^3 \end{pmatrix} &= \begin{pmatrix} \varepsilon & 0 & 0 \\ 0 & \varepsilon & 0 \\ 0 & 0 & \varepsilon_3 \end{pmatrix}\begin{pmatrix} E^1 \\ E^2 \\ E^3 \end{pmatrix} + \begin{pmatrix} \xi & 0 & 0 \\ 0 & \xi & 0 \\ 0 & 0 & \xi_3 \end{pmatrix}\begin{pmatrix} H^1 \\ H^2 \\ H^3 \end{pmatrix}, \\
\begin{pmatrix} B^1 \\ B^2 \\ B^3 \end{pmatrix} &= \begin{pmatrix} \xi & 0 & 0 \\ 0 & \xi & 0 \\ 0 & 0 & \xi_3 \end{pmatrix}\begin{pmatrix} E^1 \\ E^2 \\ E^3 \end{pmatrix} + \begin{pmatrix} \mu & 0 & 0 \\ 0 & \mu & 0 \\ 0 & 0 & \mu_3 \end{pmatrix}\begin{pmatrix} H^1 \\ H^2 \\ H^3 \end{pmatrix}
\end{aligned} \tag{32.43}$$

relative to a crystallographic basis $\{\mathbf{e}_i\}$.

33. Balance Principles

From Lorentz's formula (32.9) we see that the force acting on a moving charge due to a field of magnetic induction is perpendicular to the velocity field. Hence the power density on the moving charge is produced entirely by the electric field. On any domain $\mathscr{D}$ with current density $\mathbf{j}$ the total power is given by the integral

$$\int_{\mathscr{D}} \mathbf{j} \cdot \mathbf{E}\, dx. \tag{33.1}$$

This power represents the rate of conversion of the electromagnetic energy into other forms of energy such as thermal energy (e.g., the Joule heat). Assuming that energy is balanced in $\mathscr{D}$, we equate the power with a rate of decrease of electromagnetic energy in $\mathscr{D}$ together with the energy flux through the boundary $\partial\mathscr{D}$. Such a balance principle was considered first by Poynting.

To obtain an expression for the power (33.1) in terms of the electromagnetic field in $\mathscr{D}$, we use Maxwell's equation (32.7d) to determine the current density $\mathbf{j}$:

$$\int_{\mathscr{D}} \mathbf{j} \cdot \mathbf{E}\, dx = \frac{1}{4\pi} \int_{\mathscr{D}} \left[c\mathbf{E} \cdot \operatorname{curl} \mathbf{H} - \mathbf{E} \cdot \frac{\partial \mathbf{D}}{\partial t} \right] dx. \tag{33.2}$$

Now using the vector identity

$$\operatorname{div}(\mathbf{E} \times \mathbf{H}) = \mathbf{H} \cdot \operatorname{curl} \mathbf{E} - \mathbf{E} \cdot \operatorname{curl} \mathbf{H} \tag{33.3}$$

together with the field equation (32.7b), we can rewrite (33.2) as

$$\int_{\mathscr{D}} \mathbf{j} \cdot \mathbf{E}\, dx = -\frac{1}{4\pi} \int_{\mathscr{D}} \left[c \operatorname{div}(\mathbf{E} \times \mathbf{H}) + \mathbf{E} \cdot \frac{\partial \mathbf{D}}{\partial t} + \mathbf{H} \cdot \frac{\partial \mathbf{B}}{\partial t} \right] dx. \tag{33.4}$$

Applying the divergence theorem to the first term on the right-hand side, we obtain

$$\begin{aligned} \int_{\mathscr{D}} \mathbf{j} \cdot \mathbf{E}\, dx = &-\frac{1}{4\pi} \int_{\mathscr{D}} \left[\mathbf{E} \cdot \frac{\partial \mathbf{D}}{\partial t} + \mathbf{H} \cdot \frac{\partial \mathbf{B}}{\partial t} \right] dx \\ &- \frac{c}{4\pi} \int_{\partial\mathscr{D}} (\mathbf{E} \times \mathbf{H}) \cdot \mathbf{n}\, d\sigma, \end{aligned} \tag{33.5}$$

where $\mathbf{n}$ denotes the outward unit normal on $\partial\mathscr{D}$.

Poynting observed that the volume integral on the right-hand side may be regarded as being the rate of decrease of the energy of the electromagnetic field in $\mathscr{D}$, while the surface integral may be regarded as being the energy flux through $\partial\mathscr{D}$. Indeed, when **D** and **B** are related to **E** and **H** by some linear constitutive equations, we may define an electromagnetic field energy of the domain $\mathscr{D}$ by the integral

$$\frac{1}{8\pi}\int_{\mathscr{D}}(\mathbf{E}\cdot\mathbf{D}+\mathbf{H}\cdot\mathbf{B})\,dx. \tag{33.6}$$

This expression for energy may be justified in electromagnetostatics. For time-dependent fields we simply define the field energy by (33.6). Then (33.5) may be rewritten as

$$-\frac{d}{dt}\left[\frac{1}{8\pi}\int_{\mathscr{D}}(\mathbf{E}\cdot\mathbf{D}+\mathbf{H}\cdot\mathbf{B})\,dx\right]=\int_{\mathscr{D}}\mathbf{j}\cdot\mathbf{E}\,dx+\int_{\partial\mathscr{D}}\mathbf{S}\cdot\mathbf{n}\,d\sigma, \tag{33.7}$$

where **S** is defined by

$$\mathbf{S}=\frac{c}{4\pi}\,\mathbf{E}\times\mathbf{H} \tag{33.8}$$

and is called the *Poynting vector*.

Poynting regarded **S** as being the energy flux associated with the electromagnetic fields. Thus (33.7) may be interpreted as a balance principle which asserts that the rate of decrease of the field energy in $\mathscr{D}$ is equal to the rate of conversion of energy in $\mathscr{D}$ plus the rate of energy efflux through $\partial\mathscr{D}$. We call this assertion *Poynting's principle*. Since it is valid for all domains $\mathscr{D}$, from (33.4) or from (32.7) directly we obtain its field equation

$$\frac{1}{4\pi}\left(\mathbf{E}\cdot\frac{\partial\mathbf{D}}{\partial t}+\mathbf{H}\cdot\frac{\partial\mathbf{B}}{\partial t}\right)+\mathbf{j}\cdot\mathbf{E}+\operatorname{div}\mathbf{S}=0, \tag{33.9}$$

which is known as *Poynting's equation*.

Now it should be pointed out that Poynting's principle is really an identity which is satisfied by all solutions of Maxwell's equations. The relation between Poynting's equation and Maxwell's equations is similar to that between the energy equation (i.e., the Jacobi integral) and Lagrange's equations in analytical mechanics. Poynting's principle is not a new axiom for electromagnetism; it is merely a theorem in the context of Maxwell's equations. The definition (33.8) for the energy flux is just a convenient choice. Clearly we may add an arbitrary divergence-free term on the right-hand side of (33.8) without affecting the identity (33.7).

We can derive a balance principle for the linear momentum similar to Poynting's principle. The principle may be stated for a homogeneous bianisotropic medium in general. We regard the Lorentz force on the charge and the current in $\mathscr{D}$ as a rate of conversion of the electromagnetic field momentum into the mechanical momentum. Then this rate must be balanced by a rate of decrease of the electromagnetic field momentum in $\mathscr{D}$ together with the linear momentum flux through $\partial\mathscr{D}$. Following the procedure as before, we write the momentum conversion rate in $\mathscr{D}$ by the integral

$$\int_{\mathscr{D}} \left(q\mathbf{E} + \frac{\mathbf{j}}{c} \times \mathbf{B} \right) dx. \tag{33.10}$$

Now using Maxwell's equations (32.7a) and (32.7d) to determine the charge density q and the current density $\mathbf{j}$, we get

$$\int_{\mathscr{D}} \left(q\mathbf{E} + \frac{\mathbf{j}}{c} \times \mathbf{B} \right) dx = \frac{1}{4\pi} \int_{\mathscr{D}} \left[(\operatorname{div} \mathbf{D})\mathbf{E} + (\operatorname{curl} \mathbf{H}) \times \mathbf{B} - \frac{1}{c} \frac{\partial \mathbf{D}}{\partial t} \times \mathbf{B} \right] dx. \tag{33.11}$$

We shall now rewrite the right-hand side as a sum of a rate of change of a volume integral and a surface integral.

Using the product rule and the system of field equations (32.7), we can replace the right-hand side by

$$\begin{aligned} \int_{\mathscr{D}} \left(q\mathbf{E} + \frac{\mathbf{j}}{c} \times \mathbf{B} \right) dx &= -\frac{d}{dt} \int_{\mathscr{D}} \frac{1}{4\pi c} (\mathbf{D} \times \mathbf{B})\, dx \\ &\quad + \frac{1}{4\pi} \int_{\mathscr{D}} [(\operatorname{div} \mathbf{D})\mathbf{E} + (\operatorname{curl} \mathbf{H}) \times \mathbf{B} - \mathbf{D} \times (\operatorname{curl} \mathbf{E})]\, dx. \end{aligned} \tag{33.12}$$

The integrand of the second term on the right-hand side is the divergence of the *Maxwell stress tensor* $\mathbf{T}$, which is defined by

$$\mathbf{T} = \frac{1}{4\pi} [\tfrac{1}{2}(\mathbf{E} \cdot \mathbf{D} + \mathbf{H} \cdot \mathbf{B})\mathbf{I} - \mathbf{E} \otimes \mathbf{D} - \mathbf{H} \otimes \mathbf{B}]. \tag{33.13}$$

We can verify the formula

$$-\operatorname{div} \mathbf{T} = (\operatorname{div} \mathbf{D})\mathbf{E} + (\operatorname{div} \mathbf{B})\mathbf{H} - \mathbf{B} \times \operatorname{curl} \mathbf{H} - \mathbf{D} \times \operatorname{curl} \mathbf{E} \tag{33.14}$$

by direct calculation based on the assumption that $\boldsymbol{\epsilon}$, $\boldsymbol{\xi}$, and $\boldsymbol{\mu}$ are independent of the position $\mathbf{x}$. Substituting (33.14) into (33.12) and using the divergence theorem, we obtain

$$-\frac{d}{dt}\int_{\mathscr{D}}\frac{1}{4\pi c}(\mathbf{D}\times\mathbf{B})\,dx=\int_{\mathscr{D}}\left(q\mathbf{E}+\frac{\mathbf{j}}{c}\times\mathbf{B}\right)dx+\int_{\partial\mathscr{D}}\mathbf{Tn}\,d\sigma. \quad (33.15)$$

This identity has a form similar to (33.7).

As before we regard the left-hand side of (33.15) as being the rate of decrease of the electromagnetic field momentum in $\mathscr{D}$ and the second term on the right-hand side as being the momentum efflux through $\partial\mathscr{D}$. Then (33.15) becomes a balance principle, which asserts that the rate of decrease of the field momentum in $\mathscr{D}$ is equal to the rate of momentum conversion in $\mathscr{D}$ plus the momentum efflux through $\partial\mathscr{D}$. The field equation for this balance principle is

$$\frac{1}{4\pi c}\frac{\partial}{\partial t}(\mathbf{D}\times\mathbf{B})+\left(q\mathbf{E}+\frac{\mathbf{j}}{c}\times\mathbf{B}\right)+\operatorname{div}\mathbf{T}=\mathbf{0}. \quad (33.16)$$

Like Poynting's equation (33.9), the balance equation (33.16) is really an identity which is satisfied by all solutions of Maxwell's equations (32.7) for a homogeneous bianisotropic medium. Hence this identity does not place any additional restrictions on the electromagnetic field.

It should be noted that the Maxwell stress tensor $\mathbf{T}$, as defined by (33.13), is generally not a symmetric tensor. The skew symmetric part of this tensor represents a distribution of torque on the medium by the electromagnetic fields.

In the view of Lorentz the electromagnetic fields in the interior of a material medium actually satisfy the Maxwell–Lorentz ether relations $\mathbf{E}=\mathbf{D}$ and $\mathbf{B}=\mathbf{H}$, except that the charge density and the current density must include the bound charge and the magnetization current. In other words the system of field equations always takes the form

$$\begin{aligned}&\operatorname{div}\mathbf{E}=4\pi\hat{q}, &&\operatorname{curl}\mathbf{E}=-\frac{1}{c}\frac{\partial\mathbf{B}}{\partial t},\\&\operatorname{div}\mathbf{B}=0, &&\operatorname{curl}\mathbf{B}=\frac{4\pi}{c}\mathring{\mathbf{j}}+\frac{1}{c}\frac{\partial\mathbf{E}}{\partial t},\end{aligned} \quad (33.17)$$

but $\hat{q}$ and $\mathring{\mathbf{j}}$ are now given by

$$\hat{q}=q+\frac{1}{4\pi}\operatorname{div}\mathbf{P},\qquad \mathring{\mathbf{j}}=\mathbf{j}+\frac{1}{4\pi}\frac{\partial\mathbf{P}}{\partial t}+\frac{c}{4\pi}\operatorname{curl}\mathbf{M}. \quad (33.18)$$

Clearly the system (33.17) and (33.18) is mathematically equivalent to the system (32.7).

Using the field equations (33.17), we can derive a set of balance equations similar to the set obtained before. Specifically, instead of (33.9) we now have

$$\frac{1}{4\pi}\left(\mathbf{E}\cdot\frac{\partial\mathbf{E}}{\partial t}+\mathbf{B}\cdot\frac{\partial\mathbf{B}}{\partial t}\right)+\hat{\mathbf{j}}\cdot\mathbf{E}+\operatorname{div}\hat{\mathbf{S}}=0, \tag{33.19}$$

where

$$\hat{\mathbf{S}}=\frac{c}{4\pi}\,\mathbf{E}\times\mathbf{B}, \tag{33.20}$$

and instead of (33.16) we now have

$$\frac{1}{4\pi c}\frac{\partial}{\partial t}(\mathbf{E}\times\mathbf{B})+\left(\hat{q}\mathbf{E}+\frac{\hat{\mathbf{j}}}{c}\times\mathbf{B}\right)+\operatorname{div}\hat{\mathbf{T}}=\mathbf{0}, \tag{33.21}$$

where

$$\hat{\mathbf{T}}=\frac{1}{4\pi}\left[\tfrac{1}{2}(\mathbf{E}\cdot\mathbf{E}+\mathbf{B}\cdot\mathbf{B})\mathbf{I}-\mathbf{E}\otimes\mathbf{E}-\mathbf{B}\otimes\mathbf{B}\right]. \tag{33.22}$$

Notice that $\hat{\mathbf{T}}$ is a symmetric tensor.

The theory based on the field equations (33.17) with $\hat{q}$ and $\hat{\mathbf{j}}$ given by (33.18) is known as *Lorentz's electron theory*. We shall use this theory in Section 35 to derive the governing equation for an elastic dielectric medium.

34. Electromagnetic Waves

In Section 31 we remarked that the constant c in the field equations corresponds to the speed of light (i.e., an electromagnetic wave) in a vacuum. Now we are in a position to explain this fact.

Consider a domain which has no free charge–current distribution. If the domain is a vacuum, then the electromagnetic fields in it satisfy the field equations

$$\operatorname{div}\mathbf{D}=0, \tag{34.1a}$$

$$\operatorname{curl}\mathbf{E}=-\frac{1}{c}\frac{\partial\mathbf{B}}{\partial t}, \tag{34.1b}$$

$$\operatorname{div}\mathbf{B}=0, \tag{34.1c}$$

$$\operatorname{curl}\mathbf{H}=+\frac{1}{c}\frac{\partial\mathbf{D}}{\partial t}, \tag{34.1d}$$

and the constitutive equations

$$\mathbf{D} = \mathbf{E}, \qquad \mathbf{B} = \mathbf{H}. \tag{34.2}$$

We claim that these equations require that the electric field and the magnetic field be solutions of the wave equation with wave speed c.

This fact is more or less obvious. Indeed, taking the curl of (34.1b) and using (34.1d), we obtain

$$\nabla^2\mathbf{E} = \frac{1}{c^2}\,\frac{\partial^2\mathbf{E}}{\partial t^2}. \tag{34.3}$$

Likewise, taking the curl of (34.1d) and using (34.1b), we get

$$\nabla^2\mathbf{H} = \frac{1}{c^2}\,\frac{\partial^2\mathbf{H}}{\partial t^2}. \tag{34.4}$$

These wave equations are necessary conditions on $\mathbf{E}$ and $\mathbf{H}$. A particular solution of the wave equations is a sinusoidal wave of the form

$$\begin{aligned} \mathbf{E}(\mathbf{x}, t) &= \mathbf{E}_0 \sin(k\mathbf{n}\cdot\mathbf{x} - \omega t),\\ \mathbf{H}(\mathbf{x}, t) &= \mathbf{H}_0 \sin(k\mathbf{n}\cdot\mathbf{x} - \omega t), \end{aligned} \tag{34.5}$$

where k, ω, and $\mathbf{n}$ denote the wave number, the circular frequency, and the unit vector in the direction of the wave, respectively. To find the relations among k, ω, $\mathbf{n}$, and the field vectors $\mathbf{E}$ and $\mathbf{H}$, we substitute (34.5) into the field equations (34.1), obtaining

$$\mathbf{n}\cdot\mathbf{E} = 0, \tag{34.6a}$$

$$k\mathbf{n}\times\mathbf{E} = \frac{\omega}{c}\mathbf{H}, \tag{34.6b}$$

$$\mathbf{n}\cdot\mathbf{H} = 0, \tag{34.6c}$$

$$k\mathbf{n}\times\mathbf{H} = -\frac{\omega}{c}\mathbf{E}. \tag{34.6d}$$

These equations imply directly that $\mathbf{E}$, $\mathbf{H}$, $\mathbf{n}$ are mutually orthogonal. Also' taking the cross product of (34.6a) with $\mathbf{n}$ and using (34.6d), we get

$$k\mathbf{E} = \frac{\omega^2}{kc^2}\mathbf{E}. \tag{34.7}$$

Thus for a nontrivial solution, $\mathbf{E} \neq \mathbf{0}$,

$$\frac{\omega}{k} = c, \tag{34.8}$$

which means that the wave speed is c, a result noted before from the wave equations (34.3) and (34.4). Substituting (34.8) into (34.6b) and (34.6d), we obtain

$$\mathbf{n}\times\mathbf{E} = \mathbf{H}, \tag{34.9a}$$

$$\mathbf{n}\times\mathbf{H} = -\mathbf{E}. \tag{34.9b}$$

Thus the magnitudes E and H of $\mathbf{E}$ and $\mathbf{H}$, respectively, must be equal.

Summarizing the preceding analysis, we see that (34.5) is a solution of (34.1) if and only if $\mathbf{E}_0$, $\mathbf{H}_0$, $\mathbf{n}$ are mutually orthogonal, $E_0 = H_0$, and $\omega = ck$. Such a solution is highly specialized for the field equations (34.1), of course. First, the wave (34.5) is *polarized*; i.e., the directions of $\mathbf{E}$ and $\mathbf{H}$ are fixed in space. Second, the wave is *monochromatic*; i.e., the circular frequency ω is definite. Since (34.1) is a linear system, an electromagnetic wave in the direction $\mathbf{n}$ may be an arbitrary superposition (a sum or an integral with respect to ω) of simple waves of the form (34.6). Then the directions of $\mathbf{E}$ and $\mathbf{H}$ are no longer fixed, and the frequencies may spread over a certain spectrum. The field conditions (34.6a), (34.6c) and (34.9a), (34.9b), however, remain valid, since they are invariant under any superposition.

Having described the electromagnetic fields associated with an electromagnetic wave, we now determine the radiation pressure on an absorbing surface. Suppose that the absorbing surface is a plane which is perpendicular to the direction $\mathbf{n}$ of the wave. From the Maxwell stress tensor $\mathbf{T}$ the force acting on the plane is

$$\mathbf{Tn} = \frac{1}{4\pi}\left[\tfrac{1}{2}(E^2 + H^2)\mathbf{n} - \mathbf{E}(\mathbf{E}\cdot\mathbf{n}) - \mathbf{H}(\mathbf{H}\cdot\mathbf{n})\right] = \frac{E^2}{4\pi}\mathbf{n}, \tag{34.10}$$

where we have used the conditions (34.6a), (34.6c) and (34.9a), (34.9b). Thus the pressure is

$$p = \frac{E^2}{4\pi}. \tag{34.11}$$

From (33.8) the Poynting vector is

$$\mathbf{S} = \frac{c}{4\pi}\mathbf{E}\times\mathbf{H} = \frac{cE^2}{4\pi}\mathbf{n}. \tag{34.12}$$

Thus from (34.11)

$$\mathbf{S} = cp\mathbf{n}. \tag{34.13}$$

By following essentially the same procedure, we can analyze the behavior of electromagnetic waves in a homogeneous isotropic medium.

The field equations are still given by (34.1), but the constitutive equations now read

$$\mathbf{D} = \varepsilon\mathbf{E}, \qquad \mathbf{B} = \mu\mathbf{H}. \tag{34.14}$$

As before we take the curl of (34.1b) and then use (34.1d) and (34.14), resulting in a wave equation of the form

$$\nabla^2\mathbf{E} = \frac{\mu\varepsilon}{c^2}\,\frac{\partial^2\mathbf{E}}{\partial t^2}. \tag{34.15}$$

Similarly, taking the curl of (34.1d) and using (34.1b) and (34.14), we get

$$\nabla^2\mathbf{H} = \frac{\mu\varepsilon}{c^2}\,\frac{\partial^2\mathbf{H}}{\partial t^2}. \tag{34.16}$$

In this case the wave speed v is given by

$$v = \frac{c}{(\mu\varepsilon)^{1/2}}. \tag{34.17}$$

We consider again the sinusoidal wave of the form (34.5). When we substitute (34.5) into the field equation (34.1) on the basis of the system of constitutive equations (34.14), we obtain

$$\mathbf{n}\cdot\mathbf{E} = 0, \tag{34.18a}$$

$$k\mathbf{n}\times\mathbf{E} = \frac{\omega\mu}{c}\,\mathbf{H}, \tag{34.18b}$$

$$\mathbf{n}\cdot\mathbf{H} = 0, \tag{34.18c}$$

$$k\mathbf{n}\times\mathbf{H} = -\,\frac{\omega\varepsilon}{c}\,\mathbf{E}. \tag{34.18d}$$

Thus **E**, **H**, **n** are again mutually orthogonal. Now taking the cross product of (34.18b) with **n** and from (34.18d), we have

$$k\mathbf{E} = \frac{\omega^2\mu\varepsilon}{kc^2}\,\mathbf{E}. \tag{34.19}$$

Thus in this case the wave speed is given by

$$v = \frac{\omega}{k} = \frac{c}{(\mu\varepsilon)^{1/2}}, \tag{34.20}$$

as we have already observed in (34.17). Substituting (34.19) into (34.18), we obtain

$$\mathbf{n}\times\mathbf{E} = \left(\frac{\mu}{\varepsilon}\right)^{1/2}\mathbf{H}, \qquad \mathbf{n}\times\mathbf{H} = -\left(\frac{\varepsilon}{\mu}\right)^{1/2}\mathbf{E}. \tag{34.21}$$

Thus the magnitudes of **E** and **H** are related by

$$\varepsilon^{1/2}E = \mu^{1/2}H. \tag{34.22}$$

As explained before, an electromagnetic wave in general may be an arbitrary superposition of simple waves of the form (34.5), and the conditions (34.19a), (34.19c), (34.21), and (34.22) always hold. The formula for the radiation pressure now takes the form

$$p = \frac{\varepsilon E^2}{4\pi}. \tag{34.23}$$

The behavior of electromagnetic waves in an anisotropic medium is much more complicated than that in an isotropic medium. For simplicity we shall consider only a medium which is electrically anisotropic but magnetically isotropic. In such a medium **B** is still parallel to **H**, but **D** is generally not parallel to **E**. Suppose that $\{\mathbf{e}_i\}$ is a principal basis for the permittivity tensor $\boldsymbol{\epsilon}$, and let the index i be assigned in such a way that

$$\varepsilon_1 < \varepsilon_2 < \varepsilon_3, \tag{34.24}$$

where ε_i denotes the proper number of $\boldsymbol{\epsilon}$ corresponding to the proper vector $\mathbf{e}_i$. We consider again a sinusoidal wave of the form (34.5). From the constitutive relations (30.13) and (31.19) the field **D** and the field **B** have similar forms, viz.,

$$\begin{aligned} \mathbf{D}(\mathbf{x}, t) &= \mathbf{D}_0 \sin(k\mathbf{n}\cdot\mathbf{x} - \omega t), \\ \mathbf{B}(\mathbf{x}, t) &= \mathbf{B}_0 \sin(k\mathbf{n}\cdot\mathbf{x} - \omega t), \end{aligned} \tag{34.25}$$

where

$$\mathbf{D}_0 = \boldsymbol{\epsilon}\mathbf{E}_0, \tag{34.26a}$$

$$\mathbf{B}_0 = \mu\mathbf{H}_0. \tag{34.26b}$$

Substituting (34.5) and (34.25) into Maxwell's equations (34.1), we get

$$\mathbf{n}\cdot\mathbf{D} = 0, \tag{34.27a}$$

$$\mathbf{n}\times\mathbf{E} = \frac{v}{c}\mathbf{B}, \tag{34.27b}$$

$$\mathbf{n}\cdot\mathbf{B} = 0, \tag{34.27c}$$

$$\mathbf{n}\times\mathbf{H} = -\frac{v}{c}\mathbf{D}, \tag{34.27d}$$

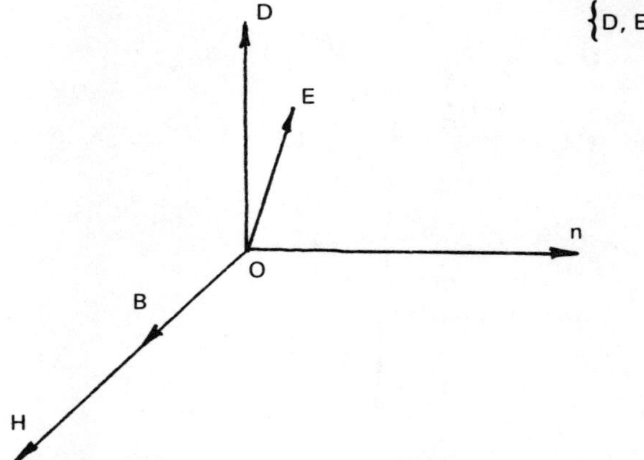

Figure 4.

where v denotes the wave speed; i.e.,

$$v = \frac{\omega}{k}. \tag{34.28}$$

The conditions (34.27a) and (34.27c) show that **D** and **B** are perpendicular to **n**. Since **H** is parallel to **B**; compared with (34.26b), the condition (34.27d) implies that **B** is perpendicular to **D**, and that the magnitudes B and D are related by

$$B = \frac{\mu v}{c} D. \tag{34.29}$$

Finally, the condition (34.27b) implies that **E** is perpendicular to **B**. Hence **E** is contained in the plane of **n** and **D**. The relative positions of the five vectors **D**, **E**, **B**, **H**, and **n** are illustrated in Fig. 4.

Now the conditions (34.27b) and (34.27d) together with the constitutive relations (34.26) form a homogeneous linear system for the pair of vectors **E** and **H**. In order that this system may have nonzero solutions, v must satisfy a propagation condition which may be derived in the following way: First, taking the cross product of (34.27b) with **n** and using (34.26), we get

$$\mathbf{E} - (\mathbf{n} \cdot \mathbf{E})\mathbf{n} = \frac{\mu v^2}{c^2} \mathbf{D}. \tag{34.30}$$

This equation may be solved for **D** in terms of **n**,

$$\left(\boldsymbol{\epsilon}^{-1} - \frac{\mu v^2}{c^2} \mathbf{I}\right)\mathbf{D} = (\mathbf{n} \cdot \mathbf{E})\mathbf{n} \tag{34.31}$$

or, equivalently,

$$\mathbf{D} = (\mathbf{n} \cdot \mathbf{E})\left(\boldsymbol{\epsilon}^{-1} - \frac{\mu v^2}{c^2}\mathbf{I}\right)^{-1}\mathbf{n}. \tag{34.32}$$

In component form relative to the principal basis $\{\mathbf{e}_i\}$ of $\boldsymbol{\epsilon}$ the equation (34.32) reads

$$D^i = \frac{c^2}{\mu}(\mathbf{n} \cdot \mathbf{E})\frac{n^i}{v_i^2 - v^2}, \tag{34.33}$$

where v_i, $i = 1, 2, 3$, are defined by

$$v_i = \frac{c}{(\varepsilon_i \mu)^{1/2}}. \tag{34.34}$$

Having obtained $\mathbf{D}$ in terms of $\mathbf{n}$, we can express the other three vectors $\mathbf{E}$, $\mathbf{B}$, and $\mathbf{H}$ by

$$\mathbf{E} = \boldsymbol{\epsilon}^{-1}\mathbf{D}, \qquad \mathbf{B} = -\frac{\mu v}{c}\mathbf{n}\times\mathbf{D}, \qquad \mathbf{H} = -\frac{v}{c}\mathbf{n}\times\mathbf{D}. \tag{34.35}$$

From (34.32) and (34.35) we verify easily that the conditions (34.27b)–(34.27d) are satisfied identically. The only remaining condition is (34.27a), which implies that

$$\frac{c^2}{\mu}(\mathbf{n} \cdot \mathbf{E})\left(\frac{(n^1)^2}{v_1^2 - v^2} + \frac{(n^2)^2}{v_2^2 - v^2} + \frac{(n^3)^2}{v_3^2 - v^2}\right) = 0. \tag{34.36}$$

This is the desired propagation condition which governs the wave speed v.

We consider three distinct cases.

(i) When n^1, n^2, n^3 all differ from zero; i.e., $\mathbf{n}$ does not belong to any coordinate plane of the principal basis $\{\mathbf{e}_i\}$. In this case $\mathbf{n} \cdot \mathbf{E}$ cannot vanish, since otherwise $\mathbf{E}$ would have to be parallel to $\mathbf{D}$, and then $\mathbf{n}$ would be perpendicular to one of the basis vectors in $\{\mathbf{e}_i\}$, contradicting the assumption that n^1, n^2, n^3 do not vanish. Thus (34.36) reduces to

$$\frac{(n^1)^2}{v_1^2 - v^2} + \frac{(n^2)^2}{v_2^2 - v^2} + \frac{(n^3)^2}{v_3^2 - v^2} = 0. \tag{34.37}$$

From the assumption (34.24) and the definition (34.34) we see that v_1, v_2, v_3 are ordered as follows:

$$v_1 > v_2 > v_3. \tag{34.38}$$

Using (34.38) and checking the changes of signs of the left-hand side of

(34.37) at the singularities v_1, v_2, v_3, we see that there are precisely two roots v^* and v_* for (34.37) such that

$$v_1 > v^* > v_2 > v_* > v_3. \tag{34.39}$$

We denote the fields associated with v^* by $\mathbf{D}^*$, $\mathbf{E}^*$, $\mathbf{B}^*$, $\mathbf{H}^*$, and those associated with v_* by $\mathbf{D}_*$, $\mathbf{E}_*$, $\mathbf{B}_*$, $\mathbf{H}_*$. From (34.32) and (34.35) we see that the directions of these fields are fixed; i.e., the waves must be polarized in a definite way.

We claim that $\mathbf{D}^*$ is perpendicular to $\mathbf{D}_*$. This fact follows directly from the formula (34.33), which implies that

$$D^{*i} = \frac{c^2}{\mu}(\mathbf{n}\cdot\mathbf{E}^*)\frac{n^i}{v_i^2 - v^{*2}}, \qquad D_*{}^i = \frac{c^2}{\mu}(\mathbf{n}\cdot\mathbf{E}_*)\frac{n^i}{v_i^2 - v_*{}^2}. \tag{34.40}$$

Hence

$$\begin{aligned}\mathbf{D}^*\cdot\mathbf{D}_* &= \frac{c^4}{\mu^2}(\mathbf{n}\cdot\mathbf{E}^*)(\mathbf{n}\cdot\mathbf{E}_*)\Bigg[\frac{(n^1)^2}{(v_1{}^2 - v^{*2})(v_1{}^2 - v_*{}^2)} \\ &\quad + \frac{(n^2)^2}{(v_2{}^2 - v^{*2})(v_2{}^2 - v_*{}^2)} + \frac{(n^3)^2}{(v_3{}^2 - v^{*2})(v_3{}^2 - v_*{}^2)}\Bigg] \\ &= \frac{c^4}{\mu^2}\,\frac{(\mathbf{n}\cdot\mathbf{E}^*)(\mathbf{n}\cdot\mathbf{E}_*)}{v^{*2} - v_*{}^2}\Bigg[\left(\frac{(n^1)^2}{v_1{}^2 - v^{*2}} + \frac{(n^2)^2}{v_2{}^2 - v^{*2}} + \frac{(n^3)^2}{v_3{}^2 - v^{*2}}\right) \\ &\quad - \left(\frac{(n^1)^2}{v_1{}^2 - v_*{}^2} + \frac{(n^2)^2}{v_2{}^2 - v_*{}^2} + \frac{(n^3)^2}{v_3{}^2 - v_*{}^2}\right)\Bigg].\end{aligned} \tag{34.41}$$

Thus $\mathbf{D}^*\cdot\mathbf{D}_*$ vanishes by virtue of the fact that v^* and v_* are roots of (34.37).

Because v_* is less than v^*, when a light beam with direction $\mathbf{n}$ exits from the boundary of the anisotropic medium and enters into an isotropic medium the diffracted beam splits[(2)] into two beams with different angles of diffraction. Each of the diffracted beams is polarized in a definite way. Similarly when a light beam exits from an isotropic medium and enters into an anisotropic medium, the diffracted beam also splits into two beams, and each of the diffracted beams is polarized in a definite way. The property of double diffraction is known as *birefringence*. The fact that the diffracted beams are polarized is used for producing polarized light, e.g., the Nicol prism.

(2) The reader is assumed to be familiar with Snell's law of diffraction.

(ii) When $\mathbf{n}$ belongs to a coordinate plane of $\{\mathbf{e}_i\}$ but does not coincide with any of the basis vectors $\mathbf{e}_i$.

There are two possibilities: First, $\mathbf{n} \cdot \mathbf{E} \neq 0$. In this case the equation (34.36) reduces to

$$\begin{aligned} \frac{(n^1)^2}{v_1{}^2 - v^2} + \frac{(n^2)^2}{v_2{}^2 - v^2} &= 0, \qquad \text{when } n^3 = 0 \\ \frac{(n^1)^2}{v_1{}^2 - v^2} + \frac{(n^3)^2}{v_3{}^2 - v^2} &= 0, \qquad \text{when } n^2 = 0 \\ \frac{(n^2)^2}{v_2{}^2 - v^2} + \frac{(n^3)^2}{v_3{}^2 - v^2} &= 0, \qquad \text{when } n^1 = 0. \end{aligned} \tag{34.42}$$

Any one of these equations has a single root v^* such that

$$\begin{aligned} v_1 > v^* > v_2, &\qquad \text{when } n^3 = 0, \\ v_1 > v^* > v_3, &\qquad \text{when } n^2 = 0, \\ v_2 > v^* > v_3, &\qquad \text{when } n^1 = 0. \end{aligned} \tag{34.43}$$

The fields $\mathbf{D}^*$, $\mathbf{E}^*$, $\mathbf{B}^*$, $\mathbf{H}^*$ associated with this wave speed are given by the representations (34.32) and (34.35). It is important that v^* be different from any one of v_1, v_2, v_3. This condition is automatically satisfied in (34.43a) and (34.43c). In (34.43b) we must have

$$v^* \neq v_2. \tag{34.44}$$

Second, $\mathbf{n} \cdot \mathbf{E} = 0$. In this case $\mathbf{E}$ must be parallel to $\mathbf{D}$. Thus $\mathbf{D}$ is a proper vector of $\boldsymbol{\epsilon}$. From Fig. 4 we see clearly that $\mathbf{D}$ must be parallel to $\mathbf{e}_i$ when $n^i = 0$. The corresponding wave speed v_* may be obtained from (34.31), which reduces to

$$\left(\boldsymbol{\epsilon}^{-1} - \frac{\mu v^2}{c^2}\,\mathbf{I}\right)\mathbf{D} = \mathbf{0} \tag{34.45}$$

in this case. Hence when $\mathbf{D}$ is parallel to $\mathbf{e}_i$, the speed v_* is just v_i, viz.,

$$v_* = \frac{c}{(\varepsilon_i \mu)^{1/2}} = v_i. \tag{34.46}$$

We denote the fields associated with this wave speed by $\mathbf{D}_*$, $\mathbf{E}_*$, $\mathbf{B}_*$, $\mathbf{H}_*$.

The preceding analysis shows that for each direction $\mathbf{n}$ there are again two distinct wave speeds v^* and v_*, and the waves associated with these speeds are polarized in a definite way. From (34.43) and (34.46) we have the order

$$\begin{aligned} &v_1 > v^* > v_2 > v_* = v_3, \qquad \text{when } n^3 = 0 \\ &v_1 = v_* > v_2 > v^* > v_3, \qquad \text{when } n^1 = 0. \end{aligned} \tag{34.47}$$

In the case $n^2 = 0$ both v^* and v_* belong to the interval (v_1, v_3). There are two exceptional cases when $v^* = v_*$.

These two cases occur in the directions $\mathbf{n}^*$ and $\mathbf{n}_*$ in the coordinate plane of $\mathbf{e}_1$ and $\mathbf{e}_3$ such that

$$\frac{(n^{*1})^2}{v_1^{\,2} - v_2^{\,2}} + \frac{(n^{*3})^2}{v_3^{\,2} - v_2^{\,2}} = 0, \tag{34.48}$$

and

$$n^{*1} = n_*^{\,1} > 0, \qquad n^{*3} = -n_*^{\,3} > 0. \tag{34.49}$$

In these two directions we have

$$v^* = v_* = v_2, \tag{34.50}$$

and the corresponding waves are no longer polarized. We call these two directions the *optic axes*. When $n^2 = 0$, but $\mathbf{n}$ is not an optic axis, the order of v_* and v^* is given by

$$\begin{aligned} &v_1 > v_2 = v_* > v^* > v_3, \qquad \text{when } |\, n^1 \,| > n^{*1} \\ &v_1 > v^* > v_2 = v_* > v_3, \qquad \text{when } |\, n^1 \,| < n^{*1}. \end{aligned} \tag{34.51}$$

(iii) When $\mathbf{n}$ coincides with one of the basis vectors $\mathbf{e}_1$, $\mathbf{e}_2$, $\mathbf{e}_3$.

In this case the equation (34.36) may be satisfied only when $\mathbf{n} \cdot \mathbf{E} = 0$. Hence as before $\mathbf{D}$ must be a proper vector of $\boldsymbol{\epsilon}$. There are three possibilities: First, $\mathbf{n} = \mathbf{e}_1$. In this case $v = v_2$ corresponds to $\mathbf{D}$ parallel to $\mathbf{e}_2$, and $v = v_3$ corresponds to $\mathbf{D}$ parallel to $\mathbf{e}_3$. The other two possibilities with $\mathbf{n} = \mathbf{e}_2$ or $\mathbf{n} = \mathbf{e}_3$ are similar. The two waves propagating in each one of these directions are polarized, and their corresponding two fields $\mathbf{D}$ are perpendicular to each other as in case (i).

A similar but somewhat simpler analysis may be given for the case that $\boldsymbol{\epsilon}$ has only two distinct proper numbers. For this case there is a single optic axis in the direction of the proper vector of $\boldsymbol{\epsilon}$ corresponding to the simple proper number.

35. Electromechanical Interactions

Up to this point in our formulation of the classical theory of electromagnetism we have allowed the free-charge distribution and the free-current distribution as well as the electromagnetic fields associated with them to vary in space and time. We have allowed the medium to interact with the electromagnetic fields through the polarization field and the magnetization field. We have not allowed the medium to move in space, however.

Physically, of course, an unrestrained conducting wire situated in a magnetic field will move when a current goes through it, since a force is generated on the wire by the field. In fact, electric motors and many other electric devices are based precisely on such induced forces. In these applications the electromechanical interactions may be explained adequately by using external supplies; i.e., the force generated by the electromagnetic field may be regarded as an external force for the mechanical problem. We need not introduce interactions among the fluxes such as stresses and electromotive forces.

The simple interactions through external supplies are not adequate to explain such physical phenomena as piezoelectric effects and photoelastic effects, however. To account for these phenomena we generalize the theory by allowing interactions not only through external supplies but also through constitutive equations. Thus motions and deformations may affect the electromagnetic properties of the medium,[3] and conversely electromagnetic fields may affect the mechanical response of the medium.[3] Such interactions are much more difficult to analyze mathematically, since the electromagnetic field equations and the mechanical field equations must be treated as a coupled system.

In this section we shall summarize a particular model for the electromechanical interactions. This model is developed by Toupin[4] for a nonmagnetic elastic dielectric medium. The electromechanical constitutive equations are obtained by using an energy principle similar to that of a hyperelastic material; cf. Section 20. A survey report on the electro-magneto-mechanical interactions has been prepared recently by Pao.[5] Several

[3] Here the medium must be a material body. Physically, it does not make too much sense to talk about a moving and deforming vacuum.

[4] R. A. Toupin, A dynamic theory of elastic dielectrics, *International Journal of Engineering Science*, Vol. 1, pp. 101–126, 1963.

[5] Y. H. Pao, *Electromagnetic Forces in Deformable Media*, Report No. 2508, Material Science Center, Cornell University, Ithaca, New York, 1975.

formulations of Maxwell's equations for moving and deforming media and models of field-matter interactions are summarized in that report.

We assume first that the dielectric medium is at rest in the ether frame. Since we are primarily interested in the interactions of the polarization and the deformation, we assume also that the free charge and the free current both vanish. Under these assumptions we may rewrite the field equation (32.7a) as

$$\operatorname{div} \mathbf{E} = -\operatorname{div} \mathbf{P} = 4\pi q_p, \tag{35.1}$$

where

$$q_p = -\frac{1}{4\pi} \operatorname{div} \mathbf{P}. \tag{35.2}$$

At the boundary of the dielectric medium the field $\mathbf{P}$ generally suffers a jump discontinuity. Then by applying the divergence theorem to the integral of (35.2) on an elementary cylindrical domain bridging the two sides of the boundary surface, we have

$$\tau_p = -\frac{1}{4\pi} [\![\mathbf{P}]\!]\mathbf{n}. \tag{35.3}$$

We may regard q_p and τ_p as the volume density and the surface density of the polarization charge distributions in the medium. These densities are known as the *Poisson–Kelvin equivalent charge distributions.*

If the polarization field on the stationary dielectric medium varies in time, then the rate of change of the total polarization charge in a domain $\mathscr{D}$ is given by

$$\frac{d}{dt} \int_{\mathscr{D}} q_p \, dx = -\frac{1}{4\pi} \int_{\partial\mathscr{D}} \frac{\partial \mathbf{P}}{\partial t} \cdot \mathbf{n} \, d\sigma = -\int_{\partial\mathscr{D}} \mathbf{j}_p \cdot \mathbf{n} \, d\sigma, \tag{35.4}$$

where

$$\mathbf{j}_p = \frac{1}{4\pi} \frac{\partial \mathbf{P}}{\partial t}. \tag{35.5}$$

Since $\mathscr{D}$ is arbitrary, we may regard $\mathbf{j}_p$ as the polarization current in the medium.

Now we remove the assumption that the dielectric medium is at rest in the ether frame. Let $\boldsymbol{\varkappa}$ be a reference configuration for the dielectric material medium. Then as explained in continuum mechanics, a motion may be described by the deformation $\boldsymbol{\varphi}_t$ from the reference configuration $\boldsymbol{\varkappa}$ to the instantaneous configuration $\boldsymbol{\chi}_t$. Specifically, we use (X^A) in $\boldsymbol{\varkappa}$

and (x^i) in χ_t. Then φ_t is given by the deformation functions

$$x^i = x^i(X^A, t). \tag{35.6}$$

The important kinematical quantities are the velocity field $\mathbf{v}$ and the deformation gradient field $\mathbf{F}$; these fields are given by the component forms

$$v^i = \frac{\partial x^i}{\partial t}, \qquad F_A{}^i = \frac{\partial x^i}{\partial X^A}, \tag{35.7}$$

as defined in Section 14.

At each time t the polarization vector $\mathbf{P}$ is defined on the instantaneous configuration χ_t. The field equation (35.1) is still valid at each point in space, and the instantaneous polarization charge densities q_p and τ_p are still given by (35.2) and (35.3). However, since the medium is moving, these charges give rise to some currents. We define the polarization current density $\bar{\mathbf{j}}_p$ due to the motion of q_p by

$$\bar{\mathbf{j}}_p = q_p \mathbf{v} = -\frac{1}{4\pi}(\operatorname{div}\mathbf{P})\mathbf{v}. \tag{35.8}$$

The polarization current due to the variations of $\mathbf{P}$ in time may be derived in the following way: Consider the rate of change of the total polarization charge in a moving material domain $\mathscr{D}_t = \varphi_t(\mathscr{D})$ given by

$$\frac{d}{dt}\int_{\varphi_t(\mathscr{D})} q_p\, dx = -\frac{1}{4\pi}\frac{d}{dt}\int_{\partial\varphi_t(\mathscr{D})} \mathbf{P}\cdot\mathbf{n}\, d\sigma. \tag{35.9}$$

We can calculate the time derivative on the right-hand side by first replacing the integral by an integral on the time-independent surface $\partial\mathscr{D}$ in the reference configuration $\varkappa$. As explained in Section 17 [cf. the proof of (17.21)], the transformation rule for the surface integral is

$$\int_{\partial\varphi_t(\mathscr{D})} \mathbf{P}\cdot\mathbf{n}\, d\sigma = \int_{\partial\mathscr{D}} \mathbf{P}_\varkappa \cdot \mathbf{N}\, d\Sigma, \tag{35.10}$$

where

$$\mathbf{P}_\varkappa = (\det\mathbf{F})\mathbf{F}^{-1}\mathbf{P}. \tag{35.11}$$

We call $\mathbf{P}_\varkappa$ the *referential polarization field* relative to the configuration $\varkappa$. Now substituting (35.10) into (35.9) and then differentiating under the integral sign, we get

$$-\frac{1}{4\pi}\frac{d}{dt}\int_{\partial\varphi_t(\mathscr{D})} \mathbf{P}\cdot\mathbf{n}\, d\sigma = -\frac{1}{4\pi}\int_{\partial\mathscr{D}} \frac{\partial\mathbf{P}_\varkappa}{\partial t}\cdot\mathbf{N}\, d\Sigma. \tag{35.12}$$

Finally, transforming the surface integral back to an integral on $\partial \boldsymbol{\varphi}_t(\mathscr{D})$, we obtain

$$-\frac{1}{4\pi}\int_{\partial\mathscr{D}} \frac{\partial \mathbf{P}_\varkappa}{\partial t}\cdot \mathbf{N}\, d\Sigma = -\frac{1}{4\pi}\int_{\partial \varphi_t(\mathscr{D})} \frac{d_c\mathbf{P}}{dt}\cdot \mathbf{n}\, d\sigma, \tag{35.13}$$

where

$$\frac{d_c\mathbf{P}}{dt} = (\det \mathbf{F})^{-1}\mathbf{F}\,\frac{\partial \mathbf{P}_\varkappa}{\partial t}, \tag{35.14}$$

which is called the *convected time derivative* of $\mathbf{P}$. From (35.11) this derivative is given by

$$\frac{d_c\mathbf{P}}{dt} = (\det \mathbf{F})^{-1}\mathbf{F}\,\frac{\partial}{\partial t}\,[(\det \mathbf{F})\mathbf{F}^{-1}\mathbf{P}], \tag{35.15a}$$

$$\frac{d_c\mathbf{P}}{dt} = \frac{\partial \mathbf{P}}{\partial t} + [\operatorname{grad} \mathbf{P}](\mathbf{v}) - [\operatorname{grad} \mathbf{v}](\mathbf{P}) + (\operatorname{div} \mathbf{v})\mathbf{P}, \tag{35.15b}$$

$$\frac{d_c\mathbf{P}}{dt} = \frac{\partial \mathbf{P}}{\partial t} + (\operatorname{div} \mathbf{P})\mathbf{v} + \operatorname{curl}(\mathbf{P}\times\mathbf{v}). \tag{35.15c}$$

Combining (35.9), (35.12), and (35.13), we see that

$$\frac{d}{dt}\int_{\varphi_t(\mathscr{D})} q_p\, dx = -\frac{1}{4\pi}\int_{\partial\varphi_t(\mathscr{D})} \frac{d_c\mathbf{P}}{dt}\cdot \mathbf{n}\, d\sigma. \tag{35.16}$$

Hence we define the polarization current $\hat{\mathbf{j}}_p$ due to the variation of $\mathbf{P}$ in time by

$$\hat{\mathbf{j}}_p = \frac{1}{4\pi}\,\frac{d_c\mathbf{P}}{dt}. \tag{35.17}$$

The total polarization current $\mathbf{j}_p$ is taken to be the sum of $\tilde{\mathbf{j}}_p$ and $\hat{\mathbf{j}}_p$, viz.,

$$\mathbf{j}_p = \tilde{\mathbf{j}}_p + \hat{\mathbf{j}}_p = \frac{1}{4\pi}\left[\frac{d_c\mathbf{P}}{dt} - (\operatorname{div} \mathbf{P})\mathbf{v}\right]. \tag{35.18}$$

From (35.15c) we have

$$\mathbf{j}_p = \frac{1}{4\pi}\left[\frac{\partial \mathbf{P}}{\partial t} + \operatorname{curl}(\mathbf{P}\times\mathbf{v})\right]. \tag{35.19}$$

Having obtained the equivalent charge and current distributions associated with the polarization field on a moving and deforming dielectric medium, we may obtain the governing equations by using Lorentz's electron theory; i.e., we regard the medium as being a vacuum which is loaded with a charge distribution q_p and a current distribution $\mathbf{j}_p$. Then the field **E** and the field **B** satisfy the system of equations

$$\operatorname{div} \mathbf{E} = 4\pi q_p, \tag{35.20a}$$

$$\operatorname{curl} \mathbf{E} = -\frac{1}{c}\frac{\partial \mathbf{B}}{\partial t}, \tag{35.20b}$$

$$\operatorname{div} \mathbf{B} = 0, \tag{35.20c}$$

$$\operatorname{curl} \mathbf{B} = \frac{4\pi}{c}\mathbf{j}_p + \frac{1}{c}\frac{\partial \mathbf{E}}{\partial t}. \tag{35.20d}$$

Substituting (35.19) into (35.20d) and using the basic relation (35.1), we obtain

$$\operatorname{curl}(\mathbf{B} + \mathbf{v}\times\mathbf{P}) = \frac{1}{c}\frac{\partial \mathbf{D}}{\partial t}, \tag{35.21}$$

which suggests that we may define the field **H** in the moving dielectric medium by

$$\mathbf{H} = \mathbf{B} + \mathbf{v}\times\mathbf{P}. \tag{35.22}$$

Then the field equations may be rewritten in the forms

$$\operatorname{div} \mathbf{D} = 0, \tag{35.23a}$$

$$\operatorname{curl} \mathbf{E} = -\frac{1}{c}\frac{\partial \mathbf{B}}{\partial t}, \tag{35.23b}$$

$$\operatorname{div} \mathbf{B} = 0, \tag{35.23c}$$

$$\operatorname{curl} \mathbf{H} = \frac{1}{c}\frac{\partial \mathbf{D}}{\partial t}, \tag{35.23d}$$

corresponding to Maxwell's equations with no free charge and no free current.

Notice that the relation (35.22) between **H** and **B** is no longer a special case of the linear constitutive relation (31.20), which is valid only for a particular type of nonferromagnetic medium at rest in the ether frame. Of course, when $\mathbf{v} = \mathbf{0}$, (35.22) reduces to

$$\mathbf{H} = \mathbf{B}, \tag{35.24}$$

regardless of what polarization field **P** the dielectric medium may have.

The electromagnetic field on the moving medium gives rise to the Lorentz force

$$\mathbf{f} = q_p\mathbf{E} + \mathbf{j}_p \times \mathbf{B} = -\frac{1}{4\pi}(\operatorname{div}\mathbf{P})\mathbf{Z} + \frac{1}{4\pi}\frac{d_c\mathbf{P}}{dt} \times \mathbf{B}, \tag{35.25}$$

where $\mathbf{Z}$ denotes the electromotive intensity, viz.,

$$\mathbf{Z} = \mathbf{E} + \mathbf{v} \times \mathbf{B}. \tag{35.26}$$

We can interpret $\mathbf{Z}$ as being the force acting on a unit charge moving with the medium. As explained in Section 33, the Lorentz force $\mathbf{f}$ gives the rate of conversion of the electromagnetic field momentum into the mechanical momentum. Similarly the energy supply γ to the medium is

$$\gamma = \mathbf{j}_p \cdot \mathbf{E} = \frac{1}{4\pi}\left[\frac{d_c\mathbf{P}}{dt} - (\operatorname{div}\mathbf{P})\mathbf{v}\right] \cdot \mathbf{E} = \frac{1}{4\pi}\frac{d_c\mathbf{P}}{dt} \cdot \mathbf{Z} + \mathbf{f} \cdot \mathbf{v}, \tag{35.27}$$

where $\mathbf{f}$ is given by (35.25). The quantity γ represents the rate of conversion of the electromagnetic field energy into the mechanical energy.

Other than the preceding momentum and energy conversions we allow the electromechanical interactions to take place in the constitutive relations also. First, we assume that the polarization vector is determined by the electromotive intensity $\mathbf{Z}$ and the deformation $\mathbf{F}$, viz.,

$$\mathbf{P} = \boldsymbol{\pi}(\mathbf{Z}, \mathbf{F}), \tag{35.28}$$

such that $\boldsymbol{\pi}$ is invertible with respect to $\mathbf{Z}$ for each $\mathbf{F}$, so that we have also

$$\mathbf{Z} = \boldsymbol{\zeta}(\mathbf{F}, \mathbf{P}). \tag{35.29}$$

The relation (35.28) with the property of invertibility (35.29) is a generalization of the linear constitutive relation (30.15) to a moving and deforming dielectric medium. Next, we assume that the mechanical response of the medium is hyperelastic, but because of the electromechanical interactions we require that the stress tensor $\mathbf{T}$ and the stored energy ε be determined by the deformation gradient $\mathbf{F}$ and the polarization vector $\mathbf{P}$, viz.,

$$\mathbf{T} = \mathbf{G}(\mathbf{F}, \mathbf{P}), \qquad \varepsilon = \varphi(\mathbf{F}, \mathbf{P}). \tag{35.30}$$

Note. From (35.28) and (35.29) $\mathbf{T}$ and ε are also functions of $\mathbf{F}$ and $\mathbf{Z}$. It is more convenient to express the stress and the stored energy as functions of $\mathbf{P}$ instead of $\mathbf{Z}$, since the momentum and the energy conversions are given in terms of $\mathbf{P}$, not in terms of $\mathbf{Z}$.

Under the preceding constitutive assumptions we may derive a relation between the stored energy function φ and the response functions $\mathbf{G}$ and $\boldsymbol{\zeta}$ by using an energy principle. The procedure is explained in detail in Section 20 for the purely mechanical case. Following that procedure, we assume first that the total energy in a moving and deforming material medium $\mathscr{D}_t = \boldsymbol{\varphi}_t(\mathscr{D})$ is given by the sum of the kinetic energy and the stored energy. Then the *energy principle for a hyperelastic dielectric material* requires that the rate of change of the total energy be balanced by the power of the forces as well as the energy supply, viz.,

$$\frac{d}{dt}\int_{\mathscr{D}_t} \varrho(\varepsilon + \tfrac{1}{2}v^2)\,dx = \int_{\mathscr{D}_t}\left[(\varrho\mathbf{b} + \mathbf{f})\cdot\mathbf{v} + \frac{1}{4\pi}\frac{d_c\mathbf{P}}{dt}\cdot\mathbf{Z}\right]dx + \int_{\partial\mathscr{D}_t}\mathbf{t}\cdot\mathbf{v}\,d\sigma \tag{35.31}$$

This principle generalizes the principle (20.4) from the purely mechanical case to the electromechanical case.

Cauchy's principles of balance of linear momentum and balance of moment of momentum, of course, remain the same as before, except that the body force field now includes explicitly the Lorentz force $\mathbf{f}$. By using these balance principles, we can reduce the energy principle to the energy equation

$$\varrho\dot{\varepsilon} = \operatorname{tr}(\mathbf{T}\operatorname{grad}\mathbf{v}) + \frac{1}{4\pi}\frac{d_c\mathbf{P}}{dt}\cdot\mathbf{Z}. \tag{35.32}$$

Now since it is more convenient to calculate the convected time derivative of $\mathbf{P}$ from the referential polarization vector $\mathbf{P}_\varkappa$, we write the energy equation in the referential form

$$\varrho_\varkappa\dot{\varepsilon} = \operatorname{tr}(\mathbf{T}_\varkappa{}^T\dot{\mathbf{F}}) + \mathbf{Z}_\varkappa\cdot\dot{\mathbf{P}}_\varkappa, \tag{35.33}$$

where $\mathbf{T}_\varkappa$ is the Piola–Kirchhoff stress tensor defined by (17.21), and where $\mathbf{Z}_\varkappa$ is the referential electromotive intensity defined similarly by

$$\mathbf{Z}_\varkappa = \frac{1}{4\pi}(\mathbf{F}^{-1})^T\mathbf{Z}. \tag{35.34}$$

From (35.11) we can express ε as a function of $\mathbf{F}$ and $\mathbf{P}_\varkappa$:

$$\varepsilon = \psi(\mathbf{F}, \mathbf{P}_\varkappa). \tag{35.35}$$

Then by the chain rule

$$\dot{\varepsilon} = \operatorname{tr}\left[\left(\frac{\partial\psi}{\partial\mathbf{F}}\right)^T\dot{\mathbf{F}}\right] + \frac{\partial\psi}{\partial\mathbf{P}_\varkappa}\cdot\dot{\mathbf{P}}_\varkappa. \tag{35.36}$$

Substituting (35.36) into (35.33) and requiring the equation to hold in all motions and for all polarization fields, we obtain

$$\mathbf{T}_\varkappa = \varrho_\varkappa \frac{\partial \psi}{\partial \mathbf{F}}, \qquad \mathbf{Z}_\varkappa = \varrho_\varkappa \frac{\partial \psi}{\partial \mathbf{P}_\varkappa}, \tag{35.37}$$

which shows that the stored energy function is the potential function for the response functions of the stress and the electromotive intensity.

The constitutive relations (35.30) must satisfy Noll's principle of material frame-indifference, of course. Specifically,

$$\mathbf{G}(\mathbf{QF}, \mathbf{QP}) = \mathbf{QG}(\mathbf{F}, \mathbf{P})\mathbf{Q}^T, \qquad \varphi(\mathbf{QF}, \mathbf{QP}) = \varphi(\mathbf{F}, \mathbf{P}) \tag{35.38}$$

for all rotations $\mathbf{Q}$. From (35.11) $\mathbf{P}_\varkappa$ remains unchanged when $\mathbf{F}$ and $\mathbf{P}$ are replaced by $\mathbf{QF}$ and $\mathbf{QP}$. Hence ψ satisfies

$$\psi(\mathbf{QF}, \mathbf{P}_\varkappa) = \psi(\mathbf{F}, \mathbf{P}_\varkappa). \tag{35.39}$$

Then as explained in Section 20 (cf. Noll's theorem), the stress tensor $\mathbf{T}$, given by

$$\mathbf{T} = \varrho \mathbf{F}\left(\frac{\partial \psi}{\partial \mathbf{F}}\right)^T, \tag{35.40}$$

is necessarily a symmetric tensor. This condition renders the equation of moment of momentum an identity in all motions. The equations of linear momentum, of course, remain the same as before, cf. (17.12), except that the external forces now include the Lorentz force $\mathbf{f}$, viz.,

$$\operatorname{div} \mathbf{T} + \varrho \mathbf{b} + \mathbf{f} = \varrho \mathbf{a}. \tag{35.41}$$

As far as the electromagnetic field equations are concerned, we still have Maxwell's equations (35.23), except that the constitutive equations are now given by

$$\begin{aligned} \mathbf{H} &= \mathbf{B} + \mathbf{v} \times (\mathbf{D} - \mathbf{E}), \\ \mathbf{D} &= \mathbf{E} + \boldsymbol{\pi}(\mathbf{E} + \mathbf{v} \times \mathbf{B}, \mathbf{F}). \end{aligned} \tag{35.42}$$

In particular, when $\boldsymbol{\pi}(\mathbf{Z}, \mathbf{F})$ depends linearly on $\mathbf{Z}$, say,

$$\boldsymbol{\pi}(\mathbf{Z}, \mathbf{F}) = [\boldsymbol{\xi}(\mathbf{F})](\mathbf{Z}), \tag{35.43}$$

where $\boldsymbol{\xi}(\mathbf{F})$ is a tensor depending on $\mathbf{F}$, we have a system of linear constitutive relations

$$\mathbf{H} = \boldsymbol{\alpha}\mathbf{E} + \boldsymbol{\beta}\mathbf{B}, \qquad \mathbf{D} = \boldsymbol{\gamma}\mathbf{E} + \boldsymbol{\delta}\mathbf{B}, \tag{35.44}$$

where the tensors $\boldsymbol{\alpha}$, $\boldsymbol{\beta}$, $\boldsymbol{\gamma}$, $\boldsymbol{\delta}$ depend on $\mathbf{v}$ and $\mathbf{F}$.

Because of the electromechanical interactions in the constitutive relations (35.40) and (35.42) or (35.44) the equation of linear momentum (35.41) and Maxwell's equations (35.23) must be treated as a coupled system of field equations for the moving and deforming hyperelastic dielectric medium.

Before closing this chapter we remark again that throughout the formulation of the classical theory of electromagnetism it is assumed that a particular frame on the Newtonian space–time, called the ether frame, is used as the frame of reference. It is generally agreed that the ether frame is an inertial frame as defined in Newtonian mechanics. In Section 32 we remarked that relative to an inertial frame in general the constitutive relations of a medium do not have the same forms as those relative to the ether frame, unless the inertial frame is at rest in the ether frame. This feature of the classical theory is not satisfactory, since physically there is no clear way to identify an inertial frame, much less to identify a particular inertial frame among the set of all inertial frames.

The famous experiment of Michelson and Morley finally shows directly that the preceding feature of the classical theory is not consistent with the behavior of electromagnetic waves in nature. Hence a new and better model is needed to characterize the principles governing electromagnetism. Such a model is presented in the following chapter.

It should be noted, however, that the use of Noll's principle of material frame-indifference in (35.38) is not inconsistent with the basic assumptions of the classical theory of electromagnetism, since the medium under consideration in this section is assumed to be elastic. In particular, (35.38) may be viewed as the transformation properties of the response functions $\mathbf{G}$ and $\boldsymbol{\varphi}$ under a static rigid transformation of the frame of reference. Such a transformation does preserve the constitutive relations in classical electromagnetism. An application of the same principle to the constitutive relations of media with memory effects, which we have not considered at all in this chapter, will not be consistent with classical electromagnetism, of course, as we have pointed out in Section 32.

6

Special Relativistic Theory of Electromagnetism

One of the difficulties in the classical theory of electromagnetism is the condition that the Maxwell–Lorentz ether relations $\mathbf{D} = \mathbf{E}$, $\mathbf{B} = \mathbf{H}$ for a vacuum are not invariant under a general Galilean transformation. As a result, the whole theory is formulated on the basis of a particular ether frame of reference, and some of its predictions are known to be inconsistent with experimental observations. This difficulty is removed in the theory of relativity. In this chapter we summarize first the mathematical model of the event world used in the special theory of relativity. Then we develop the theory of electromagnetism in the context of this mathematical model.

36. Newtonian, Galilean, and Ether Space–Times

In the preceding five chapters we have formulated the classical theories of mechanics, continuum mechanics, and electromagnetism in the context of a mathematical model for the event world, called the Newtonian space–time $\mathscr{E}$. We recall that $\mathscr{E}$ is the disjoint union of a family of oriented 3-dimensional Euclidean space $\{\mathscr{E}_\tau, \tau \in \mathscr{T}\}$,

$$\mathscr{E} = \bigcup_{\tau \in \mathscr{T}} \mathscr{E}_\tau, \tag{36.1}$$

where $\mathscr{T}$, the index set of the family, is itself an oriented 1-dimensional Euclidean space. We call $\mathscr{T}$ the space of Newtonian instants, and we call $\mathscr{E}_\tau$ the instantaneous physical space at the instant $\tau \in \mathscr{T}$.

A *Euclidean coordinate system* (x^i, t) on $\mathscr{E}$ is defined by a (positively oriented) isometry t from $\mathscr{T}$ to $\mathscr{R}$ and a (positively oriented) isometry (x^i) from $\mathscr{E}_\tau$ to $\mathscr{R}^3$ for each $\tau \in \mathscr{T}$. Such a coordinate system corresponds to an assignment of an initial instant τ_0 such that $t(\tau_0) = 0$, an assignment of an origin $\mathbf{o}(\tau) \in \mathscr{E}_\tau$ such that $x^i(\mathbf{o}(\tau)) = 0$, $i = 1, 2, 3$, and an assignment of a (positive) orthonormal basis $\{\mathbf{e}_i(\tau)\}$ in the instantaneous translation space $\mathscr{V}(\tau)$ such that each $\mathbf{x}(\tau) \in \mathscr{E}_\tau$ may be represented by

$$\mathbf{x}(\tau) - \mathbf{o}(\tau) = x^i(\mathbf{x}(\tau))\mathbf{e}_i(\tau); \tag{36.2}$$

cf. (1.4).

From the preceding definition we see that a change of Euclidean coordinate system from (x^i, t) to $(\bar{x}^i, \bar{t})$ is given by

$$\bar{x}^i = Q_j^{\ i}(t)x^j + \xi^i(t), \qquad \bar{t} = t + \alpha, \tag{36.3}$$

where α is a constant, and where $\xi^i(t)$ and $Q_j^{\ i}(t)$ are functions of t, such that $[Q_j^{\ i}(t)]$ is a rotation matrix for each $t \in \mathscr{R}$. We call such a coordinate transformation a *Euclidean transformation*. This transformation characterizes the collection Φ of all Euclidean coordinate systems on $\mathscr{E}$ in the sense that if any member (x^i, t) of Φ is given, then we can determine the entire collection Φ by arbitrary transformations from (x^i, t) of the form (36.3). In other words Φ is a maximal collection of coordinate systems on $\mathscr{E}$ with coordinate transformations given by (36.3).

Clearly if $\mathscr{E}$ is given by the disjoint union (36.1), then Φ can be defined uniquely and is a maximal collection with respect to (36.3). Conversely if $\mathscr{E}$ is a given set equipped with a maximal collection Φ of coordinate systems with coordinate transformations given by (36.3), then $\mathscr{E}$ can be defined uniquely as a disjoint union of the form (36.1), such that the given collection Φ becomes the collection of Euclidean coordinate systems on $\mathscr{E}$.

The preceding remark means that we can define a Newtonian space–time also as a pair, $(\mathscr{E}, \Phi)$, where $\mathscr{E}$ is a set, and where Φ is a maximal collection of (global) coordinate systems on $\mathscr{E}$ with coordinate transformations given by (36.3).

Mathematically the definition based on the pair $(\mathscr{E}, \Phi)$ is equivalent to the definition based on the disjoint union of the form (36.1). The coordinate-free definition (36.1) is more intuitive, so it is preferred when we wish to explain concepts related to Newtonian space–time. In applications, however, the definition based on the pair $(\mathscr{E}, \Phi)$ is more convenient, since all numerical calculations must be done in terms of a coordinate system. Hence it is important that we be familiar with both definitions.

Now in classical mechanics we distinguish a subcollection Ψ in Φ as the collection of *inertial coördinate systems*. These systems are defined by the inertial frames which are characterized by Newton's first law; cf. Section 4. In particular, we have shown that any two inertial coordinate systems (x^i, t) and $(\bar{x}^i, \bar{t})$ are related by a *Galilean transformation*:

$$\bar{x}^i = Q_j{}^i x^j + \xi^i t + \eta^i, \qquad \bar{t} = t + \alpha, \tag{36.4}$$

where α, η^i, ξ^i, and $[Q_j{}^i]$ are all constant, and where $[Q_j{}^i]$ is a rotation matrix. Clearly (36.4) is a special case of (36.3), but the converse is false. As before the coordinate transformation (36.4) characterizes the collection Ψ in the sense that if any member (x^i, t) of Ψ is given, then the entire collection Ψ may be specified by arbitrary transformations from (x^i, t) of the form (36.4). In other words Ψ is maximal with respect to (36.4).

Since Ψ is a subset of Φ, the pair $(\mathscr{E}, \Psi)$ is not the same as the pair $(\mathscr{E}, \Phi)$. We define formally a *Galilean space–time* as follows: a pair $(\mathscr{E}, \Psi)$, where $\mathscr{E}$ is a set, and where Ψ is a maximal collection of (global) coordinate systems on $\mathscr{E}$ with coordinate transformations given by (36.4).

As remarked before, this definition is not intuitive, so we proceed to find a more direct coordinate-free definition that is comparable to (36.1). In Section 43, IVT-2, we have defined the notion of an *affine space* by a triple $(\mathscr{A}, \mathscr{V}, f)$, where $\mathscr{A}$ is a set, $\mathscr{V}$ is a vector space, and f is a mapping

$$f: \mathscr{A} \times \mathscr{A} \to \mathscr{V} \tag{36.5}$$

such that

(a) $f(\mathbf{x}, \mathbf{y}) = f(\mathbf{x}, \mathbf{z}) + f(\mathbf{z}, \mathbf{y})$ for all $\mathbf{x}, \mathbf{y}, \mathbf{z}$ in $\mathscr{A}$;

(b) for every $\mathbf{x} \in \mathscr{A}$ and $\mathbf{v} \in \mathscr{V}$ there is a unique element $\mathbf{y} \in \mathscr{A}$ such that $f(\mathbf{x}, \mathbf{y}) = \mathbf{v}$.

We call $\mathscr{A}$ a Euclidean space if $\mathscr{V}$ is an inner product space, but we call $\mathscr{A}$ an affine space if $\mathscr{V}$ is just a vector space without any additional structure such as an inner product. As before $\mathscr{V}$ is called the translation space of $\mathscr{A}$, and f is called the point difference and may be denoted by

$$f(\mathbf{x}, \mathbf{y}) \equiv \mathbf{x} - \mathbf{y}. \tag{36.6}$$

Let $\mathscr{U}$ be a subspace of $\mathscr{V}$. Then $\mathscr{U}$ gives rise to an equivalence relation on $\mathscr{A}$ by

$$\mathbf{x} \sim \mathbf{y} \Leftrightarrow \mathbf{x} - \mathbf{y} \in \mathscr{U}. \tag{36.7}$$

Each equivalence class with respect to this equivalence relation is called

an *affine subspace parallel to* $\mathscr{U}$. Thus $\mathscr{U}$ gives rise to a decomposition of $\mathscr{A}$ into a disjoint union of parallel affine subspaces.

Now the Galilean space–time $\mathscr{E}$ may be defined directly as a Newtonian space–time of the form (36.1), such that $\mathscr{E}$ is also an affine space, and that (36.1) is a decomposition of $\mathscr{E}$ into a disjoint union of parallel affine subspaces.

To see that this more intuitive direct definition is equivalent to the previous definition, we have to prove the following: Given the pair $(\mathscr{E}, \Psi)$, there is an induced structure on $\mathscr{E}$ as stated in the direct definition, and conversely, given a structure on $\mathscr{E}$ as stated in the direct definition, there is a maximal collection Ψ with coordinate transformations given by (36.4), such that the induced structure of the pair $(\mathscr{E}, \Psi)$ coincides with the given structure on $\mathscr{E}$.

Suppose that $(\mathscr{E}, \Psi)$ is given. Then we determine an affine space structure on $\mathscr{E}$ in the following way: We choose the 4-dimensional translation space of $\mathscr{E}$ to be $\mathscr{E}$ itself with a particular point $\mathbf{z}$ singled out as the origin. To assign a vector space structure on $\mathscr{E}$, we take the sum of any two events $\mathbf{x}, \mathbf{y} \in \mathscr{E}$ to be the event $\mathbf{x} + \mathbf{y}$, such that the coordinates $(x^i(\mathbf{x} + \mathbf{y}), t(\mathbf{x} + \mathbf{y}))$ of $\mathbf{x} + \mathbf{y}$ are given by

$$x^i(\mathbf{x} + \mathbf{y}) \equiv x^i(\mathbf{x}) + x^i(\mathbf{y}) - x^i(\mathbf{z}), \tag{36.8a}$$

$$t(\mathbf{x} + \mathbf{y}) \equiv t(\mathbf{x}) + t(\mathbf{y}) - t(\mathbf{z}), \tag{36.8b}$$

where (x^i, t) is any member of Ψ. Of course, we must show that the event $\mathbf{x} + \mathbf{y}$ thus defined is independent of the choice of (x^i, t) in Ψ. This fact can be proved easily by using the coordinate transformation (36.4). Indeed, from (36.8) and (36.4)

$$\begin{aligned}
\bar{x}^i(\mathbf{x} + \mathbf{y}) &= Q_j{}^i[x^j(\mathbf{x}) + x^j(\mathbf{y}) - x^j(\mathbf{z})] + \xi^i[t(\mathbf{x}) + t(\mathbf{y}) - t(\mathbf{z})] + \eta^i \\
&= [Q_j{}^i x^j(\mathbf{x}) + \xi^i t(\mathbf{x}) + \eta^i] + [Q_j{}^i x^j(\mathbf{y}) + \xi^i t(\mathbf{y}) + \eta^i] \\
&\quad - [Q_j{}^i x^j(\mathbf{z}) + \xi^i t(\mathbf{z}) + \eta^i] \\
&= \bar{x}^i(\mathbf{x}) + \bar{x}^i(\mathbf{y}) - \bar{x}^i(\mathbf{z}), \\
\bar{t}(\mathbf{x} + \mathbf{y}) &= [t(\mathbf{x}) + t(\mathbf{y}) - t(\mathbf{z})] + \alpha \\
&= [t(\mathbf{x}) + \alpha] + [t(\mathbf{y}) + \alpha] - [t(\mathbf{z}) + \alpha] \\
&= \bar{t}(\mathbf{x}) + \bar{t}(\mathbf{y}) - \bar{t}(\mathbf{z}).
\end{aligned} \tag{36.9}$$

Thus the operation of addition is well defined.

Similarly, for any $\mathbf{x} \in \mathscr{E}$ and any $\alpha \in \mathscr{R}$ we take $\alpha\mathbf{x}$ to be the event such that the coordinates $(x^i(\alpha\mathbf{x}), t(\alpha\mathbf{x}))$ are given by

$$x^i(\alpha\mathbf{x}) \equiv \alpha x^i(\mathbf{x}) - (\alpha - 1)x^i(\mathbf{z}), \tag{36.10a}$$

$$t(\alpha\mathbf{x}) \equiv \alpha t(\mathbf{x}) - (\alpha - 1)t(\mathbf{z}), \tag{36.10b}$$

where (x^i, t) is any member of Ψ. Then by the same argument as before we verify that

$$\begin{aligned} \bar{x}^i(\alpha\mathbf{x}) &= \alpha\bar{x}^i(\mathbf{x}) - (\alpha - 1)\bar{x}^i(\mathbf{z}), \\ \bar{t}(\alpha\mathbf{x}) &= \alpha\bar{t}(\mathbf{x}) - (\alpha - 1)\bar{t}(\mathbf{z}), \end{aligned} \tag{36.11}$$

for any $(\bar{x}^i, \bar{t})$ related to (x^i, t) by (36.4). Thus the operation of scalar multiplication is also well defined.

We can verify that the axioms of a vector space are all satisfied by these operations; e.g., the particular event $\mathbf{z}$ plays the role of the null vector, viz.,

$$\mathbf{x} + \mathbf{z} = \mathbf{x} \tag{36.12}$$

for all $\mathbf{x} \in \mathscr{E}$, since according to (36.8) we have

$$\begin{aligned} x^i(\mathbf{x} + \mathbf{z}) &= x^i(\mathbf{x}) + x^i(\mathbf{z}) - x^i(\mathbf{z}) = x^i(\mathbf{x}), \\ t(\mathbf{x} + \mathbf{z}) &= t(\mathbf{x}) + t(\mathbf{z}) - t(\mathbf{z}) = t(\mathbf{x}). \end{aligned} \tag{36.13}$$

Next, we define the point difference operation f on $\mathscr{E}$ directly by (36.6), where on the right-hand side the operation $\mathbf{x} - \mathbf{y}$ is defined on the vector space $\mathscr{E}$; i.e.,

$$\mathbf{x} - \mathbf{y} = \mathbf{x} + (-\mathbf{y}). \tag{36.14}$$

Then the conditions (a) and (b) are clearly satisfied. Thus the pair $(\mathscr{E}, \Psi)$ gives rise to an affine space structure on $\mathscr{E}$.

The pair $(\mathscr{E}, \Psi)$ also gives rise to a decomposition of $\mathscr{E}$ into a disjoint union of the form (36.1). First, we define simultaneity of any pair of events $\mathbf{x}, \mathbf{y} \in \mathscr{E}$ by the condition

$$t(\mathbf{x}) = t(\mathbf{y}) \tag{36.15}$$

relative to any member (x^i, t) of Ψ. The transformation rule (36.4) shows clearly that (36.15) implies

$$\bar{t}(\mathbf{x}) = \bar{t}(\mathbf{y}) \tag{36.16}$$

for any other $(\bar{x}^i, \bar{t})$ in Ψ. Thus the instantaneous spaces $\mathscr{E}_\tau$ are defined, and $\mathscr{E}$ is given by the disjoint union (36.1). We define an oriented Euclidean space structure on each $\mathscr{E}_\tau$ by requiring (x^i) to be a positive rectangular Cartesian coordinate system. Then (36.4) shows that this structure is independent of the choice of (x^i, t) in Ψ. Thus the pair $(\mathscr{E}, \Psi)$ gives rise to a Newtonian space–time structure on $\mathscr{E}$.

Finally, we still have to prove that the instantaneous spaces $\mathscr{E}_\tau$ are parallel affine subspaces in $\mathscr{E}$. This fact is more or less obvious. First, the instantaneous space $\mathscr{E}_\zeta$ containing the particular event $\mathbf{z}$ is a subspace in $\mathscr{E}$. That is to say, if $\mathbf{x}$ and $\mathbf{y}$ are simultaneous with $\mathbf{z}$, then $\mathbf{x} + \mathbf{y}$ and $\alpha\mathbf{x}$ are also. This fact follows directly from the definitions (36.8b) and (36.10b). Next, if $\mathbf{x}$ is any event in $\mathscr{E}$, then another event $\mathbf{y}$ is simultaneous with $\mathbf{x}$ if and only if $\mathbf{x} - \mathbf{y}$ is simultaneous with $\mathbf{z}$. This fact also follows directly from (36.8b) and (36.10b). Indeed,

$$\begin{aligned} t(\mathbf{x} - \mathbf{y}) &= t(\mathbf{x} + (-\mathbf{y})) = t(\mathbf{x}) + t(-\mathbf{y}) - t(\mathbf{z}) = t(\mathbf{x}) - t(\mathbf{y}) + 2t(\mathbf{z}) - t(\mathbf{z}) \\ &= t(\mathbf{x}) - t(\mathbf{y}) + t(\mathbf{z}), \end{aligned} \tag{36.17}$$

where the right-hand side reduces to $t(\mathbf{z})$ if and only if $\mathbf{x}$ is simultaneous with $\mathbf{y}$. Thus each $\mathscr{E}_\tau$ is an affine subspace parallel to the subspace $\mathscr{E}_\zeta$.

Thus the proof of the fact that the pair $(\mathscr{E}, \Psi)$ gives rise to a structure on $\mathscr{E}$ as stated in the direct definition is complete.

Next, we show that, conversely, if $\mathscr{E}$ has the structure as stated in the direct definition, then we can define a collection Ψ, which is maximal with respect to the coordinate transformation (36.4), such that the induced structure on $\mathscr{E}$ by the pair $(\mathscr{E}, \Psi)$ coincides with the given structure on $\mathscr{E}$.

To define Ψ, we simply choose those Euclidean coordinate systems on $\mathscr{E}$ which are also Cartesian systems with respect to the affine space structure on $\mathscr{E}$. By virtue of the fact that Cartesian systems on an affine space are related to one another by affine transformations, it is clear that (36.3) reduces to (36.4) on the collection Ψ. In other words a Euclidean transformation is also an affine transformation if and only if it is a Galilean transformation. Thus the structure on $\mathscr{E}$ determines a unique maximal collection Ψ with coordinate transformations given by (36.4).

It is clear that the induced structure of the pair $(\mathscr{E}, \Psi)$ is precisely the same as the given structure. Indeed, the structure of vector space on $\mathscr{E}$ with $\mathbf{z}$ as the origin may be expressed by (36.8) and (36.10) if (x^i, t) is a Cartesian system with respect to the structure of affine space on $\mathscr{E}$. Similarly the structure of Euclidean space–time on $\mathscr{E}$ requires the coordinate system (x^i) to be a positive rectangular Cartesian system on each $\mathscr{E}_\tau$. As a result, the structure on $\mathscr{E}$ induced by the pair $(\mathscr{E}, \Psi)$ coincides with the given structure on $\mathscr{E}$.

Thus we have shown that a Galilean space–time may be defined either as a pair $(\mathscr{E}, \Psi)$ or more directly as a Newtonian space–time equipped with a structure of affine space, such that (36.1) is a decomposition of $\mathscr{E}$ into a disjoint union of parallel affine subspaces.

Note. The equivalence of these two definitions is entirely due to the fact that a Galilean transformation (36.4) corresponds exactly to a Euclidean transformation (36.3), which is also an affine transformation.

Next, in classical electromagnetism we distinguish a still smaller subcollection Ω of Φ as the collection of *ether coordinate systems*. These systems are defined by the ether frames, which are characterized by the system of ether relations

$$\mathbf{D} = \mathbf{E}, \qquad \mathbf{B} = \mathbf{H} \tag{36.18}$$

for a vacuum at rest in the frames. Any two ether coordinate systems (x^i, t) and $(\bar{x}^i, \bar{t})$ are related by a *rigid transformation*:

$$\bar{x}^i = Q_j{}^i x^j + \eta^i, \qquad \bar{t} = t + \alpha, \tag{36.19}$$

where α, η^i, and $Q_j{}^i$ are constant, and where $[Q_j{}^i]$ is a rotation matrix. Clearly $(\mathscr{E}, \Omega)$ is a maximal collection with respect to (36.19).

As before we define formally an *ether space–time* as follows: a pair $(\mathscr{E}, \Omega)$, where $\mathscr{E}$ is a set, and where Ω is a maximal collection of (global) coordinate systems on $\mathscr{E}$ with coordinate transformations given by (36.19).

Following the procedure in the analysis of Galilean space–time, we see that the preceding definition is equivalent to the following more intuitive direct definition: An ether space–time is just a Newtonian space–time $\mathscr{E}$, such that $\mathscr{E}$ is also a product space $\mathscr{S} \times \mathscr{T}$, and that (36.1) is a decomposition of $\mathscr{E}$ into a disjoint union of τ sections, $(\mathscr{S}, \tau)$, for all $\tau \in \mathscr{T}$.

The equivalence of $(\mathscr{E}, \Omega)$ with $\mathscr{S} \times \mathscr{T} = \bigcup_{\tau \in \mathscr{T}} (\mathscr{S}, \tau)$ is entirely due to the fact that a rigid transformation (36.19) corresponds precisely to a Euclidean transformation (36.3), such that the spatial coordinate transformation from (x^i) to $(\bar{x}^i)$ is independent of t. This property of (36.19) is mathematically equivalent to the requirement that $\mathscr{E}$ be a product space $\mathscr{S} \times \mathscr{T}$.

We have explained the various space–time structures used in the classical theories here, because in the theory of special relativity a conceptually different space–time structure, called a *Minkowskian space–time*, is used. It is important for us to understand the fundamental differences among the structures of a Newtonian space–time, a Galilean space–time, an ether space–time, and a Minkowskian space–time.

37. Minkowskian Space–Time

Like the three different space–times explained in the preceding section, a Minkowskian space–time may be defined either directly as a set equipped with a certain mathematical structure or indirectly as a pair of a set and a maximal collection of coordinate systems with respect to certain coordinate transformations. We give the direct definition first.

A *Minkowskian space–time* $\mathscr{E}$ is a 4-dimensional affine space such that its translation space $\mathscr{V}$ is equipped with a Minkowskian inner product σ and a Lorentzian orientation.

In Section 12, IVT-1, we defined the concept of an inner product on a vector space. Then in Section 43, IVT-2, we defined the concept of a Euclidean space, which is just an affine space with an inner product defined on its translation space. Thus a Minkowskian space–time is similar to a Euclidean space except that its translation space is equipped with a Minkowskian inner product, which is slightly different from an (ordinary) inner product, as we shall now explain.

Consider a symmetric bilinear function

$$\sigma : \mathscr{V} \times \mathscr{V} \to \mathscr{R} \tag{37.1}$$

on a vector space $\mathscr{V}$; i.e., σ is a symmetric second-order covariant tensor in $\mathscr{V}^* \otimes \mathscr{V}^*$. By a canonical isomorphism of $\mathscr{V}^* \otimes \mathscr{V}^*$ with $\mathscr{L}(\mathscr{V};\mathscr{V}^*)$, σ corresponds to a linear map

$$\mathbf{\Sigma} : \mathscr{V} \to \mathscr{V}^* \tag{37.2}$$

such that

$$\sigma(\mathbf{u}, \mathbf{v}) = \langle \mathbf{u}, \mathbf{\Sigma}(\mathbf{v}) \rangle \tag{37.3}$$

for all $\mathbf{u}, \mathbf{v} \in \mathscr{V}$. We say that σ is *nonsingular* if $\mathbf{\Sigma}$ is an isomorphism. Equivalently this condition means that $\sigma(\mathbf{u}, \mathbf{v}) = 0$ for all $\mathbf{v} \in \mathscr{V}$ if and only if $\mathbf{u} = \mathbf{0}$. For a nonsingular σ we define a symmetric bilinear function

$$\sigma^* : \mathscr{V}^* \times \mathscr{V}^* \to \mathscr{R} \tag{37.4}$$

on $\mathscr{V}^*$ by

$$\sigma^*(\mathbf{f}, \mathbf{g}) = \sigma(\mathbf{\Sigma}^{-1}(\mathbf{f}), \mathbf{\Sigma}^{-1}(\mathbf{g})). \tag{37.5}$$

Then σ^* is nonsingular and corresponds to the linear map

$$\mathbf{\Sigma}^* = \mathbf{\Sigma}^{-1}: \mathscr{V}^* \to \mathscr{V}; \tag{37.6}$$

i.e.,

$$\sigma^*(\mathbf{f}, \mathbf{g}) = \langle \mathbf{f}, \mathbf{\Sigma}^*(\mathbf{g}) \rangle \tag{37.7}$$

for all $\mathbf{f}, \mathbf{g} \in \mathscr{V}^*$.

The preceding results are similar to those associated with an (ordinary) inner product. The linear maps $\mathbf{\Sigma}$ and $\mathbf{\Sigma}^*$ correspond to the operations of lowering and raising of indices. Hence we call a nonsingular symmetric bilinear function on $\mathscr{V}$ a *pseudo-inner-product*. The main difference between a pseudo-inner-product and an (ordinary) inner product is that the former need not be positive definite while the latter must be. Thus for a pseudo-inner-product σ it is possible for $\sigma(\mathbf{u}, \mathbf{u})$ to vanish for some nonzero vectors $\mathbf{u}$.

As usual the quadratic function $\sigma(\mathbf{u}, \mathbf{u})$ determines the symmetric bilinear function uniquely by the polar identity

$$\sigma(\mathbf{u}, \mathbf{v}) = \tfrac{1}{4}[\sigma(\mathbf{u} + \mathbf{v}, \mathbf{u} + \mathbf{v}) - \sigma(\mathbf{u} - \mathbf{v}, \mathbf{u} - \mathbf{v})]. \tag{37.8}$$

We put

$$\begin{aligned} \mathscr{V}_+ &= \{\mathbf{u}: \sigma(\mathbf{u}, \mathbf{u}) > 0\} \cup \{\mathbf{0}\}, \\ \mathscr{V}_0 &= \{\mathbf{u}: \sigma(\mathbf{u}, \mathbf{u}) = 0\}, \\ \mathscr{V}_- &= \{\mathbf{u}: \sigma(\mathbf{u}, \mathbf{u}) < 0\} \cup \{\mathbf{0}\}, \end{aligned} \tag{37.9}$$

and we call a vector $\mathbf{u}$ in $\mathscr{V}_+$, $\mathscr{V}_0$, or $\mathscr{V}_-$, *spacelike*, *signal- like*, or *timelike*, respectively. These terms are chosen in the context of Minkowskian space–time, as we shall see later.

As explained in Section 12, IVT-1, an inner product may be characterized by an orthonormal basis. We now show that a similar result is valid for a pseudo-inner-product. That is, we claim that there exists a basis $\{\mathbf{e}_i\}$ such that

$$|\,\sigma(\mathbf{e}_i, \mathbf{e}_j)\,| = \delta_{ij} \tag{37.10}$$

for all $i, j = 1, \ldots, n$. Such a basis may be called an *orthonormal basis* relative to the pseudo-inner-product σ.

To see that an orthonormal basis $\{\mathbf{e}_i\}$ exists, we choose an arbitrary inner product g on $\mathscr{V}$, and we denote the operations of lowering and raising of indices associated with g by $\mathbf{G}$ and $\mathbf{G}^*$, respectively. Then the composition $\mathbf{G}^*\mathbf{\Sigma}$ is a symmetric tensor with respect to g. Indeed, the bilinear form of $\mathbf{G}^*\mathbf{\Sigma}$ associated with g is just σ; i.e.,

$$g(\mathbf{u}, \mathbf{G}^*\mathbf{\Sigma}\mathbf{v}) = \langle \mathbf{u}, \mathbf{\Sigma}\mathbf{v} \rangle = \sigma(\mathbf{u}, \mathbf{v}) \tag{37.11}$$

for all $\mathbf{u}, \mathbf{v} \in \mathscr{V}$. Hence by the spectral theorem for a symmetric tensor (cf. Section 27, IVT-1) there is a principal basis $\{\mathbf{f}_i\}$, such that

$$g(\mathbf{f}_i, \mathbf{f}_j) = \delta_{ij}, \qquad i, j = 1, \ldots, n, \tag{37.12}$$

and that

$$\mathbf{G}^*\mathbf{\Sigma}\mathbf{f}_i = \alpha_i \mathbf{f}_i, \qquad i = 1, \ldots, n, \tag{37.13}$$

where the proper numbers α_i, $i = 1, \ldots, n$, do not vanish, because $\mathbf{G}^*\mathbf{\Sigma}$ is an automorphism of $\mathscr{V}$. Substituting (37.13) into (37.11) and using (37.12), we obtain

$$\sigma(\mathbf{f}_i, \mathbf{f}_j) = g(\mathbf{f}_i, \alpha_j \mathbf{f}_j) = \alpha_j g(\mathbf{f}_i, \mathbf{f}_j) = \alpha_j \delta_{ij}, \tag{37.14}$$

where there is no summation on j. Hence if we define

$$\mathbf{e}_i = \frac{1}{|\alpha_i|^{1/2}} \mathbf{f}_i, \qquad i = 1, \ldots, n, \tag{37.15}$$

then the resulting basis $\{\mathbf{e}_i\}$ satisfies the condition (37.10).

We arrange the order of the basis vectors in such a way that $\mathbf{e}_1, \ldots, \mathbf{e}_k$ are spacelike while $\mathbf{e}_{k+1}, \ldots, \mathbf{e}_n$ are timelike, viz.

$$\sigma(\mathbf{e}_1, \mathbf{e}_1) = 1, \ldots, \sigma(\mathbf{e}_k, \mathbf{e}_k) = 1, \ \sigma(\mathbf{e}_{k+1}, \mathbf{e}_{k+1}) = -1, \ldots, \sigma(\mathbf{e}_n, \mathbf{e}_n) = -1. \tag{37.16}$$

Then we can verify easily that

$$\mathscr{U} = \operatorname{span}\{\mathbf{e}_1, \ldots, \mathbf{e}_k\} \subset \mathscr{V}_+, \qquad \mathscr{U}^\perp = \operatorname{span}\{\mathbf{e}_{k+1}, \ldots, \mathbf{e}_n\} \subset \mathscr{V}_-, \tag{37.17}$$

where $\mathscr{U}^\perp$ denotes the orthogonal complement of $\mathscr{U}$ with respect to σ; i.e.,

$$\mathscr{U}^\perp = \{\mathbf{w} \colon \sigma(\mathbf{u}, \mathbf{w}) = 0 \text{ for all } \mathbf{u} \in \mathscr{U}\}. \tag{37.18}$$

An orthonormal basis $\{\mathbf{e}_i\}$ associated with a pseudo-inner-product σ

is not unique, of course. However, the number k of spacelike vectors (and hence the number $n - k$ of timelike vectors) must be the same for all orthonormal bases. This assertion is known as the *inertia theorem of Sylvester*. To prove that theorem, we choose another orthonormal basis $\{\bar{\mathbf{e}}_i\}$ such that the first $\bar{k}$ vectors are spacelike while the remaining $n - \bar{k}$ vectors are timelike. Then as before we define

$$\bar{\mathscr{U}} = \operatorname{span}\{\bar{\mathbf{e}}_1, \ldots, \bar{\mathbf{e}}_{\bar{k}}\} \subset \mathscr{V}_+, \qquad \bar{\mathscr{U}}^\perp = \operatorname{span}\{\bar{\mathbf{e}}_{\bar{k}+1}, \ldots, \bar{\mathbf{e}}_n\} \subset \mathscr{V}_-. \tag{37.19}$$

Now for each $\mathbf{u} \in \mathscr{U}$ we define a unique decomposition

$$\mathbf{u} = \bar{\mathbf{u}} + \bar{\mathbf{u}}^\perp, \tag{37.20}$$

where $\bar{\mathbf{u}} \in \bar{\mathscr{U}}$ and $\bar{\mathbf{u}}^\perp \in \bar{\mathscr{U}}^\perp$. From that decomposition we define a linear map $\mathbf{L}: \mathscr{U} \to \bar{\mathscr{U}}$ by $\mathbf{L}(\mathbf{u}) = \bar{\mathbf{u}}$. Clearly $\mathbf{L}$ is one-to-one, since $\bar{\mathbf{u}} = \mathbf{0}$ implies $\mathbf{u} = \bar{\mathbf{u}}^\perp \in \mathscr{V}_+ \cap \mathscr{V}_-$; i.e., $\mathbf{u} = \mathbf{0}$. Thus

$$k = \dim \mathscr{U} \leqq \dim \bar{\mathscr{U}} = \bar{k}. \tag{37.21}$$

But the roles of $\mathscr{U}$ and $\bar{\mathscr{U}}$ may be reversed in the preceding analysis; we see that k must be the same as $\bar{k}$.

The number $n - k$ distinguishes a pseudo-inner-product σ from an (ordinary) inner product, so it is called the *index* of σ. When the index is zero, σ is actually an (ordinary) inner product. Now we are ready to define the concept of a Minkowskian inner product on the translation space $\mathscr{V}$ of the Minkowskian space–time $\mathscr{E}$.

A *Minkowskian inner product* σ on $\mathscr{V}$ is a pseudo-inner-product with index 1. Since $\mathscr{V}$ is 4-dimensional, an orthonormal basis[1] $\{\mathbf{e}_\alpha\}$ with respect to σ contains three spacelike vectors[1] $\{\mathbf{e}_i\}$ and one timelike vector $\mathbf{e}_4$.

Next, we define the concept of a *Lorentzian orientation* on $\mathscr{V}$. Like an (ordinary) orientation, a Lorentzian orientation may be characterized by an equivalence class of orthonormal bases of $\mathscr{V}$. We say that $\{\mathbf{e}_\alpha\}$ and $\{\bar{\mathbf{e}}_\alpha\}$ are equivalent or have the same Lorentzian orientation if they satisfy the condition

$$\det[\sigma(\mathbf{e}_i, \bar{\mathbf{e}}_j)] > 0, \qquad \sigma(\mathbf{e}_4, \bar{\mathbf{e}}_4) > 0. \tag{37.22}$$

Relative to the preceding equivalence relation the totality of orthonormal

(1) We use Einstein's notations here: Greek letters $\alpha, \beta, \gamma, \ldots$ denote indices ranging from 1 to 4, while Latin letters $i, j, k, \ldots$ denote indices ranging from 1 to 3.

bases of $\mathscr{V}$ decomposes into four equivalence classes. We distinguish one particular equivalence class among these four classes and call that class the *Lorentz class*. For any $\{\mathbf{e}_\alpha\}$ in the Lorentz class the timelike vector $\mathbf{e}_4$ is said to *point into the future*, and the set of spacelike vectors $\{\mathbf{e}_i\}$ is said to be *right handed*.

Let $\{\bar{\mathbf{e}}_\alpha\}$ be any orthonormal basis which may or may not be in the Lorentz class. Then the timelike vector $\bar{\mathbf{e}}_4$ is said to *point into the future* (respectively, *point into the past*) if $\sigma(\mathbf{e}_4, \bar{\mathbf{e}}_4) > 0$ [respectively, $\sigma(\mathbf{e}_4, \bar{\mathbf{e}}_4) < 0$], and the set of spacelike vectors $\{\bar{\mathbf{e}}_i\}$ is said to be *right handed* (respectively, *left handed*) if $\det[\sigma(\mathbf{e}_i, \bar{\mathbf{e}}_j)] > 0$ (respectively, $\det[\sigma(\mathbf{e}_i, \bar{\mathbf{e}}_j)] < 0$), where $\{\mathbf{e}_\alpha\}$ is any orthonormal basis in the Lorentz class. From (37.22) it is easy to verify that the preceding conditions are independent of the choice of $\{\mathbf{e}_\alpha\}$ in the Lorentz class.

Using the concepts just defined, we may describe the four equivalence classes of orthonormal bases having the same Lorentzian orientation as follows: (i) The Lorentz class is the equivalence class of bases $\{\mathbf{e}_\alpha\}$ such that $\mathbf{e}_4$ points into the future and that $\{\mathbf{e}_i\}$ is right handed. (ii) The equivalence class of bases $\{\tilde{\mathbf{e}}_\alpha\}$ such that $\tilde{\mathbf{e}}_4$ points into the past and that $\{\tilde{\mathbf{e}}_i\}$ is right handed. (iii) The equivalence class of bases $\{\hat{\mathbf{e}}_\alpha\}$ such that $\hat{\mathbf{e}}_4$ points into the future and that $\{\hat{\mathbf{e}}_i\}$ is left handed. (iv) The equivalence class of bases $\{\check{\mathbf{e}}_\alpha\}$ such that $\check{\mathbf{e}}_4$ points into the past and that $\{\check{\mathbf{e}}_i\}$ is left handed.

It should be noted that for any two orthonormal bases, $\{\mathbf{e}_\alpha\}$ and $\{\bar{\mathbf{e}}_\alpha\}$, the span of $\{\mathbf{e}_i\}$ generally need not coincide with the span of $\{\bar{\mathbf{e}}_i\}$, or, equivalently, the span of $\mathbf{e}_4$ need not coincide with the span of $\bar{\mathbf{e}}_4$. However, by using the linear isomorphism $\mathbf{L}$ in the proof of the inertia theorem of Sylvester, we see that $\sigma(\mathbf{e}_4, \bar{\mathbf{e}}_4) \neq 0$ and $\det[\sigma(\mathbf{e}_i, \bar{\mathbf{e}}_j)] \neq 0$. Hence the quantities $\sigma(\mathbf{e}_4, \bar{\mathbf{e}}_4)$ and $\det[\sigma(\mathbf{e}_i, \bar{\mathbf{e}}_j)]$ must be either positive or negative.

The Lorentzian orientation may be viewed as a refinement of an (ordinary) orientation. As usual we define an orientation on $\mathscr{V}$ by an equivalence class of orthonormal bases with respect to the following equivalence relation: $\{\mathbf{e}_\alpha\}$ and $\{\bar{\mathbf{e}}_\alpha\}$ are equivalent or have the same orientation if $\det[\sigma(\mathbf{e}_\alpha, \bar{\mathbf{e}}_\beta)] > 0$. Relative to this equivalence relation the totality of orthonormal bases of $\mathscr{V}$ decomposes into two equivalence classes only. In fact one equivalence class is just the union of the previous classes (i) and (iv), while the other equivalence class is just the union of the previous classes (ii) and (iii). By using the Lorentz class, we can then distinguish a particular union, namely, the union containing the Lorentz class, as the *positive class*. Hence a Lorentzian orientation gives rise to an ordinary orientation, but the converse is false, of course.

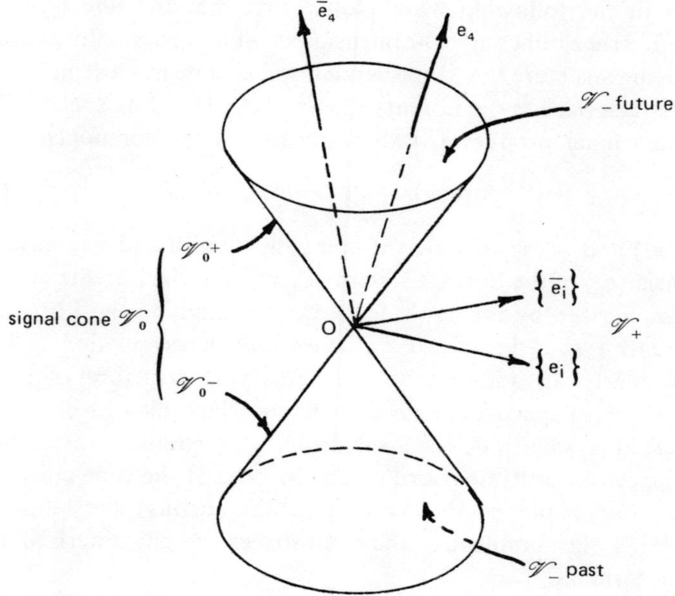

Figure 5.

The Minkowskian inner product and the Lorentzian orientation on $\mathscr{V}$ may be described schematically by Fig. 5.

So far we have discussed the direct definition of the Minkowskian space–time $\mathscr{E}$. As we have pointed out at the beginning of this section, the space–time may also be defined indirectly by a collection of coordinate systems on $\mathscr{E}$. It is required that the collection be maximal with respect to certain coordinate transformations. We proceed now to define these coordinate systems.

We call any basis $\{\mathbf{e}_\alpha\}$ in the Lorentz class a *Lorentz basis* or a *Lorentz frame*. A pair $(\mathbf{o}, \{\mathbf{e}_\alpha\})$, where $\mathbf{o} \in \mathscr{E}$ is singled out as the origin and where $\{\mathbf{e}_\alpha\}$ is a Lorentz basis, gives rise to a Cartesian coordinate system (x^i, x^4) on $\mathscr{E}$ in the usual way:

$$\mathbf{x} = \mathbf{o} + x^i\mathbf{e}_i + x^4\mathbf{e}_4, \qquad \mathbf{x} \in \mathscr{E}. \tag{37.23}$$

We call such a coordinate system on $\mathscr{E}$ a *Lorentz system*, and we denote the totality of all Lorentz systems on $\mathscr{E}$ by Θ. We shall discuss the coordinate transformations among Lorentz systems in the following section.

Using the collection Θ of Lorentz systems, we may compare the mathematical structure of a Minkowskian space–time with that of a Galilean space–time in the following way: Notice first that the underlying point sets of both space–times are 4-dimensional affine spaces. In addition to this underlying structure the Minkowskian space–time has a structure which may be characterized by a Lorentz system (x^i, x^4) in Θ. Specifically, the Minkowskian inner product σ is characterized by the component formula

$$\sigma(\mathbf{u}, \mathbf{v}) = u^i v^i - u^4 v^4, \tag{37.24}$$

where (u^i, u^4) and (v^i, v^4) denote the components of $\mathbf{u}$ and $\mathbf{v}$ relative to the Lorentz basis $\{\mathbf{e}_\alpha\}$ of the Lorentz system (x^i, x^4), and the Lorentzian orientation is characterized by the equivalence class containing the Lorentz basis of the Lorentz system in accord with the equivalence relation defined by (37.22). Similarly, in addition to the underlying structure of an affine space, the Galilean space–time has a structure which may be characterized by an inertial system (x^i, t). Specifically, the parallel (instantaneous) affine subspaces are just the coordinate subspaces of the time coordinate t, and the (ordinary) inner products on the (instantaneous) translation spaces of the parallel (instantaneous) affine subspaces are characterized by the component formula

$$\mathbf{u} \cdot \mathbf{v} = u^i v^i, \tag{37.25}$$

where u^i and v^i are the components of the (simultaneous, spatial) vectors $\mathbf{u}$ and $\mathbf{v}$ relative to the basis $\{\mathbf{e}_i\}$ of the spatial coordinate system.

In order that Minkowskian space–time may replace Galilean space–time as a model for the event world, we regard a certain Lorentz system (x^i, x^4) as a replacement of the inertial system (x^i, t) associated with the ether frame of classical electromagnetism such that x^4 corresponds to ct. In other words, in that Lorentz system the coordinates (x^α) of any event are just the list of four numbers (x^i, ct), where (x^i) are the spatial coordinates, and where t is the time of that event relative to the ether frame. Having established the relation between one particular Lorentz system and one particular inertial system, we can extend the relation to Lorentz systems and inertial systems in general by appropriate coordinate transformations.

Since the physical interpretation of an inertial frame in general or of the ether frame in particular are not really known, the Lorentz system and the inertial system in the preceding transition from a Galilean space–time to a Minkowskian space–time are also unspecified. We shall formulate the governing equations of electromagnetism in such a way that the forms of the equations are the same in all Lorentz systems. This way, the inconsistency of the classical theory with respect to the experimental result of Michelson and Morley is no longer present in the new model.

Since we cannot, and do not even wish to, identify a particular Lorentz system as being the replacement of the particular inertial system associated with the ether frame, we shall write the fourth coordinate x^4 of any Lorentz system as ct, where c denotes the speed of light in a vacuum as observed in that Lorentz system, and where t is the time coordinate in that Lorentz system. In this sense the first three coordinates x^i, $i = 1, 2, 3$, become the spatial coordinates at each time t. Since the equations of electromagnetism have the same forms in all Lorentz systems, c is a universal constant which is independent of the choice of the Lorentz system. Of course, the time coordinate t and the spatial coordinates x^i, $i = 1, 2, 3$, in a Lorentz system do not satisfy the transformation laws of the same in an inertial system. We shall consider transformations of Lorentz systems in detail in the following section.

38. Lorentz Transformations

Let (x^i, ct) be a Lorentz system on the Minkowskian space–time $\mathscr{E}$ defined by the origin $\mathbf{o} \in \mathscr{E}$ and the Lorentz basis $\{\mathbf{e}_i, \mathbf{e}_4\}$ as explained in the preceding section. We consider first a simple coordinate transformation from (x^i, ct) to another Lorentz system $(\bar{x}^i, c\bar{t})$, such that the two systems share the same origin $\mathbf{o}$ and the same two spacelike basis vectors, say,

$$\mathbf{e}_2 = \bar{\mathbf{e}}_2, \qquad \mathbf{e}_3 = \bar{\mathbf{e}}_3. \tag{38.1}$$

We claim that such a simple coordinate transformation has the explicit form

$$\bar{x}^1 = \frac{x^1 + vt}{[1 - (v/c)^2]^{1/2}}, \qquad \bar{x}^2 = x^2, \qquad \bar{x}^3 = x^3, \qquad \bar{t} = \frac{t + (vx^1/c)}{[1 - (v/c)^2]^{1/2}}, \tag{38.2}$$

where v is a constant such that $|v| < c$.

To prove that the simple coordinate transformation has the explicit form (38.2), we notice first that (38.1) and (37.10) imply the representations

$$\mathbf{e}_1 = \alpha\bar{\mathbf{e}}_1 + \beta\bar{\mathbf{e}}_4, \qquad \mathbf{e}_4 = \gamma\bar{\mathbf{e}}_1 + \lambda\bar{\mathbf{e}}_4, \tag{38.3}$$

such that the components α, β, γ, λ satisfy the conditions

$$\alpha^2 - \beta^2 = 1, \tag{38.4a}$$

$$\gamma^2 - \lambda^2 = -1, \tag{38.4b}$$

$$\alpha\gamma - \beta\lambda = 0. \tag{38.4c}$$

From (38.4c) if we put

$$\frac{\beta}{\alpha} = \frac{\gamma}{\lambda} = \frac{v}{c}, \tag{38.5}$$

where v is a parameter characterizing the change of basis, then (38.4a) and (38.4b) imply that

$$\alpha = \lambda = \frac{1}{[1 - (v/c)^2]^{1/2}}, \tag{38.6a}$$

$$\beta = \gamma = \frac{v/c}{[1 - (v/c)^2]^{1/2}}. \tag{38.6b}$$

Now using (37.23), (38.3), and (38.6), we obtain

$$\bar{x}^1 = \alpha x^1 + \gamma ct = \frac{x^1 + vt}{[1 - (v/c)^2]^{1/2}}, \qquad \bar{t} = \lambda t + \frac{\beta}{c} x^1 = \frac{t + vx^1/c^2}{[1 - (v/c)^2]^{1/2}}. \tag{38.7}$$

Thus the explicit form (38.2) is established.

In the physical interpretation the transformation (38.2) possesses several important features: First, the line with coordinates $(0, 0, 0, ct)$ in the system (x^i, ct) has coordinates $(v\bar{t}, 0, 0, c\bar{t})$ in the system $(\bar{x}^i, c\bar{t})$. Thus the origin of (x^i) corresponds to a point moving with speed v in the direction of $\bar{\mathbf{e}}_1$ as observed in the system $(\bar{x}^i, c\bar{t})$. (If v is negative, then the actual direction of motion is $-\bar{\mathbf{e}}_1$.) Next, the line with coordinates $(1, 0, 0, ct)$ in the system (x^i, ct) has coordinates $(1/[1 - (v/c)^2]^{1/2} + v\bar{t}, 0, 0, c\bar{t})$ in the system $(\bar{x}^i, c\bar{t})$. Thus the unit spatial distance as observed in the system (x^i, ct) between the lines $(0, 0, 0, ct)$ and $(1, 0, 0, ct)$ at any instant t becomes a spatial distance $1/[1 - (v/c)^2]^{1/2}$ as observed in the system $(\bar{x}^i, c\bar{t})$ between the lines $(v\bar{t}, 0, 0, c\bar{t})$ and $(1/[1 - (v/c)^2]^{1/2} + v\bar{t}, 0, 0, c\bar{t})$ at any instant $\bar{t}$. This result shows clearly that the spatial distance, calculated on the basis of the spatial coordinates (x^i) in a Lorentz system (x^i, ct), is not invariant under a change of Lorentz system. Indeed, the concept of a spatial distance is meaningful with respect to a particular Lorentz system only. Similarly the concept of a time interval is meaningful with respect to a particular system but is not invariant under a change of Lorentz system. Indeed, the unit time interval as observed in (x^i, ct) between the events $(0, 0, 0, 0)$ and $(0, 0, 0, c)$ becomes a time interval $1/[1 - (v/c)^2]^{1/2}$ as observed in the system $(\bar{x}^i, c\bar{t})$ between the events with coordinates $(0, 0, 0, 0)$ and $(v/[1 - (v/c)^2]^{1/2}, 0, 0, c/[1 - (v/c)^2]^{1/2})$. Such changes of length and time intervals are known as the *Fitzgerald–Lorentz contractions*.

In general, two Lorentz systems (x^i, ct) and $(\bar{x}^i, c\bar{t})$ need not share the same origin and the same two spacelike basis vectors, of course. Then the transformation from (x^i, ct) to $(\bar{x}^i, c\bar{t})$ may be expressed by

$$\bar{x}^i = \alpha_j{}^i x^j + \gamma^i ct + \zeta^i, \qquad c\bar{t} = \beta_i x^i + \lambda ct + \xi, \tag{38.8}$$

where $\alpha_j{}^i$, β_i, γ^i, λ are determined by the component forms of the Lorentz basis $\{\mathbf{e}_i, \mathbf{e}_4\}$ relative to the Lorentz basis $\{\bar{\mathbf{e}}_i, \bar{\mathbf{e}}_4\}$, viz.,

$$\mathbf{e}_i = \alpha_i{}^j \bar{\mathbf{e}}_j + \beta_i \bar{\mathbf{e}}_4, \qquad \mathbf{e}_4 = \gamma^i \bar{\mathbf{e}}_i + \lambda \bar{\mathbf{e}}_4. \tag{38.9}$$

Since both $\{\mathbf{e}_i, \mathbf{e}_4\}$ and $\{\bar{\mathbf{e}}_i, \bar{\mathbf{e}}_4\}$ are orthonormal bases, we must have

$$\alpha_i{}^j \alpha_k{}^j - \beta_i \beta_k = \delta_{ik}, \tag{38.10a}$$

$$\alpha_i{}^j \gamma^j - \beta_i \lambda = 0, \tag{38.10b}$$

$$\gamma^i \gamma^i - \lambda^2 = -1. \tag{38.10c}$$

Using matrix notation,[2] we can express these conditions as

$$\boldsymbol{\alpha}^T \boldsymbol{\alpha} - \boldsymbol{\beta} \otimes \boldsymbol{\beta} = \mathbf{I}, \tag{38.11a}$$

$$\boldsymbol{\alpha}^T \boldsymbol{\gamma} - \lambda \boldsymbol{\beta} = \mathbf{0}, \tag{38.11b}$$

$$\gamma^2 - \lambda^2 = -1. \tag{38.11c}$$

A general solution for (38.11) may be obtained in the following way: First, from (38.11c)

$$\lambda = (\gamma^2 + 1)^{1/2}, \tag{38.12}$$

where we have selected the positive sign for the square root because of the condition (37.22b). From (38.12) and (38.11b)

$$\boldsymbol{\beta} = \frac{1}{(\gamma^2 + 1)^{1/2}} \boldsymbol{\alpha}^T \boldsymbol{\gamma} = \frac{\gamma}{(\gamma^2 + 1)^{1/2}} \boldsymbol{\alpha}^T \mathbf{n}, \tag{38.13}$$

where $\mathbf{n} = (n^i)$ denotes the unit vector in the direction of $\boldsymbol{\gamma}$; i.e.,

$$\mathbf{n} = \frac{\boldsymbol{\gamma}}{\gamma}. \tag{38.14}$$

Substituting (38.13) into (38.11a), we obtain

$$\boldsymbol{\alpha}^T \left[\mathbf{I} - \left(1 - \frac{1}{(\gamma^2 + 1)^{1/2}} \right) \mathbf{n} \otimes \mathbf{n} \right]^2 \boldsymbol{\alpha} = \mathbf{I}, \tag{38.15}$$

[2] As usual, the subscript is the matrix column index and the superscript is the row index.

which means that the matrix

$$\boldsymbol{\eta} \equiv \left[\mathbf{I} - \left(1 - \frac{1}{(\gamma^2+1)^{1/2}}\right)\mathbf{n} \otimes \mathbf{n}\right]\boldsymbol{\alpha} \tag{38.16}$$

satisfies the condition of orthogonality, viz.,

$$\boldsymbol{\eta}^T\boldsymbol{\eta} = \mathbf{I}. \tag{38.17}$$

By using (38.16) and (37.22), we see that $\boldsymbol{\eta}$ is a rotation matrix. Then $\boldsymbol{\alpha}$ may be solved from (38.16):

$$\boldsymbol{\alpha} = \{\mathbf{I} - [1 - (\gamma^2+1)^{1/2}]\mathbf{n} \otimes \mathbf{n}\}\boldsymbol{\eta}. \tag{38.18}$$

Substituting this solution into (38.13), we obtain

$$\boldsymbol{\beta} = \gamma\boldsymbol{\eta}^T\mathbf{n}. \tag{38.19}$$

Hence the length of the vector $\boldsymbol{\beta}$ is the same as that of the vector $\boldsymbol{\gamma}$, but the direction of $\boldsymbol{\beta}$ differs from that of $\boldsymbol{\gamma}$ by the rotation $\boldsymbol{\eta}$. The formulas (38.18), (38.19), and (38.12) now give the general solution of the condition (38.11) in terms of an arbitrary nonzero vector $\boldsymbol{\gamma} = \gamma\mathbf{n}$ and an arbitrary rotation matrix $\boldsymbol{\eta}$.

It is customary to put

$$\frac{\gamma}{(\gamma^2+1)^{1/2}} = \frac{v}{c}. \tag{38.20}$$

Then we have as before

$$\gamma = \frac{v/c}{[1-(v/c)^2]^{1/2}}. \tag{38.21}$$

Hence the vector $\boldsymbol{\gamma} = (\gamma^i)$ is given by

$$\gamma^i = \frac{(v/c)n^i}{[1-(v/c)^2]^{1/2}}. \tag{38.22}$$

From (38.20) and (38.12) the scalar λ is given by

$$\lambda = \frac{1}{[1-(v/c)^2]^{1/2}}. \tag{38.23}$$

Similarly, from (38.20) and (38.18) the matrix $\boldsymbol{\alpha} = [\alpha_j{}^i]$ is given by

$$\alpha_j{}^i = \left[\delta_k{}^i - \left(1 - \frac{1}{[1-(v/c)^2]^{1/2}}\right)n^i n_k\right]\eta_j{}^k. \tag{38.24}$$

Finally, from (38.20) and (38.19) the vector $\boldsymbol{\beta} = (\beta_i)$ is given by

$$\beta_i = \frac{v/c}{[1 - (v/c)^2]^{1/2}} \eta_i{}^j n_j. \tag{38.25}$$

The formulas (38.22)–(38.25) give explicitly the coefficients of the coordinate transformation (38.8) in terms of an arbitrary rotation matrix $[\eta_j{}^i]$, an arbitrary unit vector (n^i), and an arbitrary positive number v such that $v < c$.

The set of general solutions (38.22)–(38.25) reduces to the set of simple solutions (38.6) when we choose $[\eta_j{}^i]$ to be $[\delta_j{}^i]$ and (n_i) to be $(1, 0, 0)$. Indeed, in that case (β_i) and (γ^i) reduce to

$$(\beta_i) = (\gamma^i) = \left(\frac{v/c}{[1 - (v/c)^2]^{1/2}}, 0, 0\right), \tag{38.26}$$

while $[\alpha_j{}^i]$ reduces to

$$[\alpha_j{}^i] = \begin{bmatrix} \dfrac{1}{[1 - (v/c)^2]^{1/2}} & 0 & 0 \\ 0 & 0 & 0 \\ 0 & 0 & 0 \end{bmatrix}. \tag{38.27}$$

The constant λ is, of course, always given by (38.23), which is the same as (38.6a).

A coordinate transformation of the form (38.2) with coefficients given by (38.22)–(38.25) is known as an *inhomogeneous restricted Lorentz transformation*. If the origins of (x^i, ct) and $(\bar{x}^i, c\bar{t})$ coincide, i.e., $\zeta^i = 0$, $\xi = 0$, then the transformation is known as a *(homogeneous) restricted Lorentz transformation*. If we remove the restriction (37.22), i.e., λ may be positive or negative, and $[\alpha_j{}^i]$ may be proper or improper, then the transformation is known as an *(unrestricted) Lorentz transformation*.

The inhomogeneous restricted Lorentz transformations are the desired coordinate transformations which characterize the collection Θ of all Lorentz systems (x^i, ct) on $\mathscr{E}$. Given any one Lorentz system (x^i, ct), we can determine all other Lorentz systems by arbitrary inhomogeneous restricted Lorentz transformations from (x^i, ct). In other words Θ is maximal with respect to the transformations (38.8), such that $\alpha_j{}^i$, β_i, γ^i, and λ are given by (38.24), (38.25), (38.22), and (38.23), respectively.

Since each Minkowskian inner product σ on $\mathscr{V}$ determines uniquely the maximal collection Θ, and conversely, each maximal collection Θ determines uniquely a Minkowskian inner product σ and a Lorentzian orientation on $\mathscr{V}$, we can define a Minkowskian space–time either by the direct definition given at the beginning of the preceding section or by the indirect definition based on the maximal collection Θ.

Mathematically, a Galilean transformation (36.4) may be regarded as the limit of an inhomogeneous restricted Lorentz transformation when $c \to \infty$. In this sense a Galilean space–time may be regarded as a limiting case of a Minkowskian space–time. The main difference between these two space–times is that in the translation space of the underlying 4-dimensional affine spaces, the former is equipped with a distinguished 3-dimensional subspace $\mathscr{V}_\zeta$, while the latter is equipped with a distinguished cone $\mathscr{V}_0$, which separates the timelike vectors from the spacelike vectors. A maximal subspace consisting of spacelike vectors must be 3-dimensional, but such a subspace is not unique. As a result, an instantaneous physical space is no longer assigned at each event, but we may define a 3-dimensional space relative to a Lorentz frame $\{\mathbf{e}_i, \mathbf{e}_4\}$.

To implement Minkowskian space–time as a mathematical model for the event world, we have to assign to each event the Lorentz coordinates (x^i, ct) associated with one particular Lorentz system in Θ. A procedure for making such an assignment has been explained in detail by Synge[(3)]. Since we are primarily interested in the mathematical structure of the relativistic model, we shall not discuss this procedure here.

39. Vectors and Tensors in the Minkowskian Space–Time

Since a Lorentz transformation in general does not preserve the subspace spanned by the spacelike basis vectors $\{\mathbf{e}_i\}$ of a Lorentz frame $\{\mathbf{e}_i, \mathbf{e}_4\}$, a vector of the spatial form

$$\mathbf{p} = p^i \mathbf{e}_i \tag{39.1}$$

relative to $\{\mathbf{e}_i, \mathbf{e}_4\}$ will not be of the same form relative to another Lorentz frame $\{\bar{\mathbf{e}}_i, \bar{\mathbf{e}}_4\}$. Indeed, if $\{\mathbf{e}_i, \mathbf{e}_4\}$ and $\{\bar{\mathbf{e}}_i, \bar{\mathbf{e}}_4\}$ are related by (38.9), then

$$\mathbf{p} = p^i \alpha_i{}^j \bar{\mathbf{e}}_j + p^i \beta_i \bar{\mathbf{e}}_4, \tag{39.2}$$

where $\alpha_j{}^i$ and β_i are given by (38.24) and (38.25), respectively. Thus it is meaningless to regard $\mathbf{p}$ as a purely spatial vector. We can only say that $\mathbf{p}$

(3) J. L. Synge, *Relativity: The Special Theory*, North Holland Publishing Co., Amsterdam, 1956.

is spacelike; i.e.,

$$\sigma(\mathbf{p}, \mathbf{p}) = p^i p^i = p^i \alpha_i{}^j p^k \alpha_k{}^j - p^i \beta_i p^j \beta_j > 0, \tag{39.3}$$

but generally such a vector may have nonzero components in both the spacelike basis vectors as well as in the timelike basis vector of a Lorentz frame. In other words, $p^4 = 0$ relative to one Lorentz frame $\{\mathbf{e}_i, \mathbf{e}_4\}$ by no means implies that $\bar{p}^4 = 0$ relative to another Lorentz frame $\{\bar{\mathbf{e}}_i, \bar{\mathbf{e}}_4\}$.

In the preceding section we denoted a change of Lorentz frames by

$$\mathbf{e}_\mu = e_\mu{}^\nu \bar{\mathbf{e}}_\nu, \tag{39.4}$$

where $\mu = 1, \ldots, 4$, and where ν is summed from 1 to 4 such that $[e_\mu{}^\nu]$ has the explicit form

$$[\mathbf{e}] = \left[\begin{array}{c|c} \boldsymbol{\alpha}^T & \boldsymbol{\gamma} \\ \hline \boldsymbol{\beta} & \lambda \end{array}\right]. \tag{39.5}$$

The set of conditions (38.10) on $\boldsymbol{\alpha}$, $\boldsymbol{\beta}$, $\boldsymbol{\gamma}$ and λ is equivalent to the requirement that

$$\Sigma_{\nu\varrho} e_\mu{}^\nu e_\tau{}^\varrho = \Sigma_{\mu\tau}, \tag{39.6}$$

where $[\Sigma_{\mu\nu}]$ denotes the component matrix of the Minkowskian inner product; i.e.,

$$[\Sigma_{\mu\nu}] = \operatorname{diag}(1, 1, 1, -1). \tag{39.7}$$

Using the formula (39.4) for the change of basis, we can express the transformation law of the components of a vector $\mathbf{u}$ in general by

$$\bar{u}^\nu = e_\mu{}^\nu u^\mu. \tag{39.8}$$

Similarly, the transformation law of the components of a tensor $\mathbf{A}$ of order r is

$$\bar{A}^{\nu_1 \cdots \nu_r} = e^{\nu_1}_{\mu_1} \cdots e^{\nu_r}_{\mu_r} A^{\mu_1 \cdots \mu_r}. \tag{39.9}$$

The operations of lowering and raising of indices associated with the Minkowskian inner product σ allow us to express the components in covariant form or in mixed forms also; e.g.,

$$u_\mu = \Sigma_{\mu\nu} u^\nu, \qquad A^{\mu_1}{}_{\mu_2}{}^{\mu_3 \cdots \mu_r} \equiv \Sigma_{\mu_2\nu_2} A^{\mu_1 \nu_2 \mu_3 \cdots \mu_r}, \qquad \text{etc.} \tag{39.10}$$

Then

$$\bar{u}_\nu = \bar{e}_\nu{}^\mu u_\mu, \qquad \bar{A}^{\nu_1}{}_{\nu_2}{}^{\nu_3 \cdots \nu_r} = e^{\nu_1}_{\mu_1} \bar{e}^{\mu_2}_{\nu_2} e^{\nu_3}_{\mu_3} \cdots e^{\nu_r}_{\mu_r} A^{\mu_1}{}_{\mu_2}{}^{\mu_3 \cdots \mu_r}, \qquad \text{etc.,} \tag{39.11}$$

where $[\bar{e}_\nu{}^\mu]$ denotes the σ inverse of $[e_\nu{}^\mu]$; i.e.,

$$[\bar{\mathbf{e}}] = [\mathbf{\Sigma}]\left[\begin{array}{c|c} \boldsymbol{\alpha} & \boldsymbol{\beta} \\ \hline \boldsymbol{\gamma} & \lambda \end{array}\right][\mathbf{\Sigma}] = \left[\begin{array}{c|c} \boldsymbol{\alpha} & -\boldsymbol{\beta} \\ \hline -\boldsymbol{\gamma} & \lambda \end{array}\right], \tag{39.12}$$

which satisfies the condition

$$[\mathbf{e}][\bar{\mathbf{e}}] = [\mathbf{\Sigma}]. \tag{39.13}$$

The preceding equation is equivalent to

$$\left[\begin{array}{c|c} \boldsymbol{\alpha}^T & -\boldsymbol{\gamma} \\ \hline -\boldsymbol{\beta} & \lambda \end{array}\right][\mathbf{\Sigma}]\left[\begin{array}{c|c} \boldsymbol{\alpha} & -\boldsymbol{\beta} \\ \hline -\boldsymbol{\gamma} & \lambda \end{array}\right] = [\bar{\mathbf{\Sigma}}] = [\mathbf{\Sigma}], \tag{39.14}$$

which means that $[\Sigma_{\mu\nu}]$ is the component matrix of a tensor, namely, the Minkowskian inner product σ, of course.

To illustrate the concept of a vector in Minkowskian space–time, we consider the relativistic kinematics of a particle. A motion of the particle may be described by a curve $\mathbf{w}(s) \in \mathscr{E}$ such that the tangent vector $\dot{\mathbf{w}}(s)$ is a timelike vector for all values of s. Such a curve may be reparametrized, if necessary, so as to satisfy the condition

$$\sigma(\dot{\mathbf{w}}, \dot{\mathbf{w}}) = -c^2. \tag{39.15}$$

When this condition of normalization holds, we call $\mathbf{w}(s)$ a *world line*, and we call the parameter s the *proper time*.

Now since $\dot{\mathbf{w}}(s)$ is a timelike vector at $\mathbf{w}(s)$, it may be expressed in component form relative to a Lorentz frame $\{\mathbf{e}_i, \mathbf{e}_4\}$,

$$\dot{\mathbf{w}}(s) = \dot{x}^i(s)\mathbf{e}_i + c\dot{t}(s)\mathbf{e}_4, \tag{39.16}$$

where (x^i, ct) denotes the Lorentz system defined by the Lorentz frame. In that system we can express the spatial coordinates $x^i(s)$ as functions of the time coordinate $t(s)$, viz.,

$$x^i = \varphi^i(t) = x^i(s(t)), \tag{39.17}$$

where $s(t)$ denotes the inverse of the function $t(s)$; i.e.,

$$t(s(t)) \equiv t, \qquad s(t(s)) \equiv s. \tag{39.18}$$

Then we can define an ordinary velocity vector $\mathbf{v}$ as usual by

$$\mathbf{v} \equiv \frac{d\varphi^i}{dt}\mathbf{e}_i = \frac{\dot{x}^i}{\dot{t}}\mathbf{e}_i = v^i\mathbf{e}_i, \tag{39.19}$$

where the superposed dot refers to the derivative with respect to the proper time s, as shown in (39.16). From (39.19) we see that the ordinary speed relative to the frame $\{\mathbf{e}_i, \mathbf{e}_4\}$ is

$$v = (\dot{x}^i\dot{x}^i)^{1/2}/\dot{t}. \tag{39.20}$$

Substituting the component form (39.16) into the condition of normalization of (39.15), we see that

$$\dot{x}^i\dot{x}^i - c^2\dot{t}^2 = -c^2. \tag{39.21}$$

Hence from (39.20) and (39.21) $\dot{t}$ is given by

$$\dot{t} = \frac{1}{[1 - (v/c)^2]^{1/2}}. \tag{39.22}$$

We denote the spatial component of $\dot{\mathbf{w}}$ relative to $\{\mathbf{e}_i, \mathbf{e}_4\}$ by $\mathbf{u}$, and we call $\mathbf{u}$ the *relativistic velocity* of the point as observed in the Lorentz system (x^i, ct). From (39.19) and (39.22) $\mathbf{u}$ is related to $\mathbf{v}$ by

$$\mathbf{u} = \dot{x}^i\mathbf{e}_i = \dot{t}\mathbf{v} = \frac{v^i}{[1 - (v/c)^2]^{1/2}}\,\mathbf{e}_i. \tag{39.23}$$

Numerically the components of $\mathbf{u}$ relative to the spatial basis $\{\mathbf{e}_i\}$ differ from those of $\mathbf{v}$ by a factor $\dot{t}$, which is nearly equal to 1, when v is small compared to the speed of light c.

As explained in analytical mechanics, we assign a positive number m, called the *proper mass*, to a particle. Then we define the *momentum-energy vector* $\mathbf{l}$ by

$$\mathbf{l} = m\dot{\mathbf{w}}. \tag{39.24}$$

We call the spatial component $m\dot{x}^i\mathbf{e}_i = m\mathbf{u}$ of $\mathbf{l}$ the *relativistic momentum* and the timelike component $mc^2\dot{t}^2\mathbf{e}_4$ of $\mathbf{l}$ the *relativistic energy* of the particle relative to the Lorentz system (x^i, ct).

Having considered the relativistic kinematics of a particle, we formulate next the equations of motion. As we explained at the end of Section 37, when we replace Galilean space–time by Minkowskian space–time as a model for the event world, an inertial system is replaced by a Lorentz system. Hence the forms of the relativistic equations of motion must be the same in all Lorentz systems. Moreover, when the speed of the particle relative to a Lorentz system is small compared to the speed of light, the equations of motion in that system must be approximately the same as the classical Newton's equations. From experiments of the motions of electrons in a β ray, Lorentz suggested that the relativistic vector equation of motion may be expressed as

$$\dot{\mathbf{l}} = \frac{d}{ds}\left(m\frac{d\mathbf{w}}{ds}\right) = \boldsymbol{\xi}, \tag{39.25}$$

where $\boldsymbol{\xi}$ is a vector defined at each event $\mathbf{w}(s)$ of the world line of the particle. We call $\boldsymbol{\xi}$ the *momentum-energy supply vector*.

It should be noted that, since $\dot{\mathbf{w}}$ satisfies the condition of normalization (39.15), and since m is a constant, the value of $\sigma(\mathbf{l}, \mathbf{l})$ is a constant, $-mc^2$, independent of the proper time s. As a result, the momentum-energy supply vector $\boldsymbol{\xi}$ must be orthogonal to the timelike tangent vector $\dot{\mathbf{w}}$ of the world line of the particle at each event $\mathbf{w}(s)$. In particular, $\boldsymbol{\xi}$ must be a spacelike vector.

Since (39.25) is a vector equation expressed directly in the Minkowskian space–time $\mathscr{E}$, it does not depend on the choice of any Lorentz system. Of course, we can take the component form of (39.25) relative to a Lorentz system (x^i, ct):

$$m\ddot{x}^i = \xi^i, \qquad mc\ddot{t} = \xi^4. \tag{39.26}$$

Then the orthogonality condition $\sigma(\dot{\mathbf{w}}, \boldsymbol{\xi}) = 0$ implies that

$$\dot{x}^i\xi^i - c\dot{t}\xi^4 = 0. \tag{39.27}$$

We may express the equation of motion (39.25) in terms of the ordinary velocity $v = v^i\mathbf{e}_i$ relative to a Lorentz system (x^i, ct) also. The spatial components of (39.25) correspond to

$$\frac{d}{dt}(mu^i) = \frac{d}{dt}\left(\frac{mv^i}{[1-(v/c)^2]^{1/2}}\right) = \frac{\xi^i}{\dot{t}} = f^i, \qquad i = 1, 2, 3, \tag{39.28}$$

where the vector $\mathbf{f} = f^i\mathbf{e}_i$ may be called the *relativistic force* as observed in the Lorentz system. The equations (39.28) may be regarded as the relativistic corrections to Newton's equations of motion (4.3) in the sense that the Lorentz system (x^i, ct) corresponds to an inertial system (x^i, t), and that the relativistic momentum $m\mathbf{u}$ replaces the ordinary momentum $m\mathbf{v}$.

Next, the timelike component of the equation of motion (39.25) may be written as

$$\frac{d}{dt}\left(\frac{mc^2}{[1-(v/c)^2]^{1/2}}\right) = v^i f^i, \tag{39.29}$$

where we have used the condition of orthogonality (39.27). Clearly the right-hand side of (39.29) is just the power of the force $\mathbf{f}$. Hence the quantity

$mc^2/[1 - (v/c)^2]^{1/2}$ may be called the *relativistic energy*. Assuming that v is small compared to c, we expand the energy as a power series

$$\frac{mc^2}{[1-(v/c)^2]^{1/2}} = mc^2\left[1 + \frac{1}{2}\left(\frac{v}{c}\right)^2 + \frac{3}{8}\left(\frac{v}{c}\right)^4 + \cdots\right]$$
$$= mc^2 + \frac{1}{2}mv^2 + \cdots. \tag{39.30}$$

Since mc^2 is a constant, which may be called the *rest energy* or the *proper energy* of the point, the equation (39.29) is given approximately by

$$\frac{d}{dt}\left(\frac{1}{2}mv^2 + \cdots\right) = v^i f^i. \tag{39.31}$$

Thus (39.29) is just the relativistic version of the energy equation.

From the preceding interpretation of the equations (39.26) we see that the concepts of momentum and energy are not separable in the theory of relativity, since they are the components of a single vector $\mathbf{l}$ relative to a Lorentz frame. By the same token the concepts of force (i.e., supply of momentum) and power (i.e., supply of energy) are not separable, and they are the components of a single vector $\boldsymbol{\xi}$, which is orthogonal to the tangent vector $\dot{\mathbf{w}}$ at all events of the world line of the particle.

The most important feature of the relativistic equation of motion (39.25) is that the equation is entirely independent of the choice of the Lorentz system. Hence the component forms (39.26) or (39.28) and (39.29) are invariant under any Lorentz transformation. In other words if the coordinates $w^\alpha(s)$ of $\mathbf{w}$ in one Lorentz system (x^α) satisfy the set of equations (39.26), then the coordinates $\bar{w}^\alpha(s)$ of $\mathbf{w}$ in any other Lorentz system $(\bar{x}^\alpha)$ satisfy a set of equations similar to (39.26) with ξ^α replaced by $\bar{\xi}^\alpha$. We have attained this invariance property by using an equation which is expressed directly in terms of some vectors in the Minkowskian space–time $\mathscr{E}$. In the following section we shall follow a similar approach to formulate the governing system of field equations in the special relativistic theory of electromagnetism.

40. Maxwell's Equations in Special Relativistic Form

In the preceding sections we have formulated the mathematical structure of the Minkowskian space–time $\mathscr{E}$. Now we apply that structure to electromagnetism. As we shall see, Maxwell's equations correspond to two tensor equations which are independent of the choice of any Lorentz frame. Hence the component equations are automatically invariant under any restricted Lorentz transformation. The importance of such a transformation property has been remarked before in the context of the relativistic equation of motion (39.25), which has the invariant component form (39.26) relative to any Lorentz frame $\{\mathbf{e}_i, \mathbf{e}_4\}$.

To motivate the relativistic form of Maxwell's equations, we recall first that a tensor equation or a vector equation in the 4-dimensional space $\mathscr{V}$ generally correspond to a system of tensor equations or a system of vector equations with respect to the decomposition of $\mathscr{V}$ into a 3-dimensional space span$\{\mathbf{e}_i\}$ and a 1-dimensional space span$\{\mathbf{e}_4\}$ associated with any Lorentz frame $\{\mathbf{e}_i, \mathbf{e}_4\}$. For instance the vector equation (39.25) corresponds to the system of momentum equations and energy equation as shown in (39.26) relative to any Lorentz frame. For the equations in electromagnetism the tensor equations are expressed in terms of certain differential forms.

Note. General concepts concerning differential forms are explained in detail in Sections 51 and 70, IVT-2.

Consider a 2-form $\mathbf{A}$ in the Minkowskian space–time $\mathscr{E}$. In component form relative to the dual basis $\{\mathbf{e}^i, \mathbf{e}^4\}$ of a Lorentz frame $\{\mathbf{e}_i, \mathbf{e}_4\}$, $\mathbf{A}$ is given by

$$\mathbf{A} = (P^1\mathbf{e}^1 + P^2\mathbf{e}^2 + P^3\mathbf{e}^3) \wedge \mathbf{e}^4 + (Q^1\mathbf{e}^2 \wedge \mathbf{e}^3 + Q^2\mathbf{e}^3 \wedge \mathbf{e}^1 + Q^3\mathbf{e}^1 \wedge \mathbf{e}^2), \tag{40.1}$$

where P^i and Q^i are functions of the coordinates (x^i, ct). Suppose that $\mathbf{A}$ is a closed 2-form; i.e., $\mathbf{A}$ satisfies the tensor equation

$$d\mathbf{A} = \mathbf{0}. \tag{40.2}$$

Then from (40.1)

$$\begin{aligned} d\mathbf{A} = \Bigg[&\left(\frac{\partial P^2}{\partial x^1} - \frac{\partial P^1}{\partial x^2} + \frac{1}{c}\,\frac{\partial Q^3}{\partial t}\right)\mathbf{e}^1 \wedge \mathbf{e}^2 \\ &+ \left(\frac{\partial P^3}{\partial x^2} - \frac{\partial P^2}{\partial x^3} + \frac{1}{c}\,\frac{\partial Q^1}{\partial t}\right)\mathbf{e}^2 \wedge \mathbf{e}^3 \\ &+ \left(\frac{\partial P^1}{\partial x^3} - \frac{\partial P^3}{\partial x^1} + \frac{1}{c}\,\frac{\partial Q^2}{\partial t}\right)\mathbf{e}^3 \wedge \mathbf{e}^1\Bigg] \wedge \mathbf{e}^4 \\ &+ \left(\frac{\partial Q^1}{\partial x^1} + \frac{\partial Q^2}{\partial x^2} + \frac{\partial Q^3}{\partial x^3}\right)\mathbf{e}^1 \wedge \mathbf{e}^2 \wedge \mathbf{e}^3. \end{aligned} \tag{40.3}$$

We recognize immediately that (40.3) corresponds to a system of two tensor equations:

$$\operatorname{curl} \mathbf{P} + \frac{1}{c}\frac{\partial \mathbf{Q}}{\partial t} = \mathbf{0}, \qquad \operatorname{div} \mathbf{Q} = 0, \tag{40.4}$$

with respect to the decomposition of $\mathscr{V}$ induced by the frame $\{\mathbf{e}_i, \mathbf{e}_4\}$.

Similarly, suppose that $\mathbf{A}$ satisfies the more general tensor equation

$$d\mathbf{A} = \mathbf{C}, \tag{40.5}$$

which means that the 3-form $\mathbf{C}$ is exact, and that $\mathbf{A}$ is its potential. Let the component form of $\mathbf{C}$ relative to $\{\mathbf{e}^i, \mathbf{e}^4\}$ be

$$\mathbf{C} = (R^1\mathbf{e}^2 \wedge \mathbf{e}^3 + R^2\mathbf{e}^3 \wedge \mathbf{e}^1 + R^3\mathbf{e}^1 \wedge \mathbf{e}^2) \wedge \mathbf{e}^4 + S\mathbf{e}^1 \wedge \mathbf{e}^2 \wedge \mathbf{e}^3. \tag{40.6}$$

Then from (40.3) the tensor equation (40.5) corresponds to a system of two tensor equations

$$\operatorname{curl} \mathbf{P} + \frac{1}{c}\frac{\partial \mathbf{Q}}{\partial t} = \mathbf{R}, \qquad \operatorname{div} \mathbf{Q} = S, \tag{40.7}$$

with respect to the same decomposition induced by $\{\mathbf{e}_i, \mathbf{e}_4\}$.

Comparing (40.4) and (40.7) with (32.7), we see that the tensor equations (40.2) and (40.5) are just what we need to express Maxwell's equations in relativistic form. Specifically, relative to any Lorentz frame $\{\mathbf{e}_i, \mathbf{e}_4\}$ we define the *relativistic electromagnetic field* $\mathbf{\Phi}$ by

$$\mathbf{\Phi} = (E^1\mathbf{e}^1 + E^2\mathbf{e}^2 + E^3\mathbf{e}^3) \wedge \mathbf{e}^4 + (B^1\mathbf{e}^2 \wedge \mathbf{e}^3 + B^2\mathbf{e}^3 \wedge \mathbf{e}^1 + B^3\mathbf{e}^1 \wedge \mathbf{e}^2), \tag{40.8}$$

where E^i and B^i are regarded as the components of the field $\mathbf{E}$ and the field $\mathbf{B}$, respectively, relative to the spatial (ether) frame $\{\mathbf{e}_i\}$. Then the governing equation for $\mathbf{\Phi}$ is the tensor equation

$$d\mathbf{\Phi} = \mathbf{0}, \tag{40.9}$$

which corresponds to the system

$$\operatorname{curl} \mathbf{E} + \frac{1}{c}\frac{\partial \mathbf{B}}{\partial t} = \mathbf{0}, \qquad \operatorname{div} \mathbf{B} = 0. \tag{40.10}$$

Since $\mathscr{E}$ is retractible, by Poincaré's lemma (cf. Section 52, IVT-2) the closed 2-form $\mathbf{\Phi}$ is exact; i.e., there exists a 1-form $\mathbf{\Pi}$, called the *relativ-*

istic electromagnetic potential, such that

$$\mathbf{\Phi} = d\mathbf{\Pi}. \tag{40.11}$$

Writing the 1-form $\mathbf{\Pi}$ in component form relative to $\{\mathbf{e}^i, \mathbf{e}^4\}$, we have

$$\mathbf{\Pi} = A^1\mathbf{e}^1 + A^2\mathbf{e}^2 + A^3\mathbf{e}^3 + \zeta\mathbf{e}^4, \tag{40.12}$$

where A^i and ζ are functions of the coordinates (x^i, ct). Taking the exterior derivative of (40.12), we get

$$\begin{aligned} d\mathbf{\Pi} = &\left[\left(\frac{\partial\zeta}{\partial x^1} - \frac{1}{c}\,\frac{\partial A^1}{\partial t}\right)\mathbf{e}^1 + \left(\frac{\partial\zeta}{\partial x^2} - \frac{1}{c}\,\frac{\partial A^2}{\partial t}\right)\mathbf{e}^2 \right. \\ &\left. + \left(\frac{\partial\zeta}{\partial x^3} - \frac{1}{c}\,\frac{\partial A^3}{\partial t}\right)\mathbf{e}^3\right] \wedge \mathbf{e}^4 + \left[\left(\frac{\partial A^3}{\partial x^2} - \frac{\partial A^2}{\partial x^3}\right)\mathbf{e}^2 \wedge \mathbf{e}^3 \right. \\ &\left. + \left(\frac{\partial A^1}{\partial x^3} - \frac{\partial A^3}{\partial x^1}\right)\mathbf{e}^3 \wedge \mathbf{e}^1 + \left(\frac{\partial A^2}{\partial x^1} - \frac{\partial A^1}{\partial x^2}\right)\mathbf{e}^1 \wedge \mathbf{e}^2\right]. \end{aligned} \tag{40.13}$$

Hence the tensor equation (40.11) corresponds to the system

$$\mathbf{E} = \operatorname{grad} \zeta - \frac{1}{c}\,\frac{\partial \mathbf{A}}{\partial t}, \qquad \mathbf{B} = \operatorname{curl} \mathbf{A}, \tag{40.14}$$

which are the potential equations for the electromagnetic fields.

Next, we define the *relativistic charge–current field* $\mathbf{\Psi}$ by

$$\mathbf{\Psi} = \left(\frac{j^1}{c}\,\mathbf{e}^2 \wedge \mathbf{e}^3 + \frac{j^2}{c}\,\mathbf{e}^3 \wedge \mathbf{e}^1 + \frac{j^3}{c}\,\mathbf{e}^1 \wedge \mathbf{e}^2\right) \wedge \mathbf{e}^4 - q\mathbf{e}^1 \wedge \mathbf{e}^2 \wedge \mathbf{e}^3, \tag{40.15}$$

where j^i and q are regarded as the components of the field $\mathbf{j}$ and the field q, respectively, relative to the spatial (ether) frame $\{\mathbf{e}_i\}$. Similarly, we define the *relativistic charge–current potential* $\mathbf{\Gamma}$ by

$$\mathbf{\Gamma} = (H^1\mathbf{e}^1 + H^2\mathbf{e}^2 + H^3\mathbf{e}^3) \wedge \mathbf{e}^4 - (D^1\mathbf{e}^2 \wedge \mathbf{e}^3 + D^2\mathbf{e}^3 \wedge \mathbf{e}^1 + D^3\mathbf{e}^1 \wedge \mathbf{e}^2), \tag{40.16}$$

where H^i and D^i are regarded as the components of the field $\mathbf{H}$ and the field $\mathbf{D}$, respectively, relative to the spatial (ether) frame $\{\mathbf{e}_i\}$ as before. Then the law governing the charge–current field $\mathbf{\Psi}$ is

$$d\mathbf{\Gamma} = 4\pi\mathbf{\Psi}, \tag{40.17}$$

which is a tensor equation in the space–time $\mathscr{E}$ independent of any Lorentz frame. From (40.7) the tensor equation (40.17) corresponds to the system

$$\operatorname{curl} \mathbf{H} = 4\pi \frac{\mathbf{j}}{c} + \frac{1}{c} \frac{\partial \mathbf{D}}{\partial t}, \qquad \operatorname{div} \mathbf{D} = 4\pi q \tag{40.18}$$

relative to the decomposition induced by the Lorentz frame $\{\mathbf{e}_i, \mathbf{e}_4\}$.

The condition that $\boldsymbol{\Psi}$ is an exact 3-form implies that

$$d\boldsymbol{\Psi} = \mathbf{0}. \tag{40.19}$$

Taking the exterior derivative of $\boldsymbol{\Psi}$ from (40.15), we see that the tensor equation (40.19) corresponds precisely to the continuity equation (32.1). Thus (40.19) characterizes the law of conservation of electric charge in relativistic form.

It should be emphasized again that the relativistic laws of electromagnetism (40.9), (40.11), (40.17), and (40.19) are stated directly in terms of the differential forms. Consequently, these laws are entirely independent of any choice of Lorentz frame. In particular, the components of the forms $\boldsymbol{\Phi}$, $\boldsymbol{\Pi}$, $\boldsymbol{\Psi}$ and $\boldsymbol{\Gamma}$ relative to the Lorentz systems in the collection Θ must satisfy the transformation law (39.11). As a result, the transformation laws for the fields **E**, **H**, **B**, **D**, **j**, q under a Lorentz transformation are no longer at our disposal. Indeed, we can derive the relativistic transformation laws for the components E^i, H^i, B^i, D^i, j^i, and q under a Lorentz transformation from the general formula (39.11).

For simplicity we consider the special Lorentz transformation (38.2) only. In that transformation $\{\mathbf{e}^i, \mathbf{e}^4\}$ and $\{\bar{\mathbf{e}}^i, \bar{\mathbf{e}}^4\}$ are related by

$$\bar{\mathbf{e}}^1 = \frac{\mathbf{e}^1 + (v/c)\mathbf{e}^4}{[1 - (v/c)^4]^{1/2}}, \qquad \bar{\mathbf{e}}^2 = \mathbf{e}^2, \qquad \bar{\mathbf{e}}^3 = \mathbf{e}^3, \qquad \bar{\mathbf{e}}^4 = \frac{\mathbf{e}^4 + (v/c)\mathbf{e}^1}{[1 - (v/c)^2]^{1/2}}. \tag{40.20}$$

Substituting (40.20) into the component form of $\boldsymbol{\Phi}$ in the $(\bar{x}^i, c\bar{t})$ system, viz.,

$$\begin{aligned}
\boldsymbol{\Phi} &= (\bar{E}^1\bar{\mathbf{e}}^1 + \bar{E}^2\bar{\mathbf{e}}^2 + \bar{E}^3\bar{\mathbf{e}}^3) \wedge \bar{\mathbf{e}}^4 + (\bar{B}^1\bar{\mathbf{e}}^2 \wedge \bar{\mathbf{e}}^3 + \bar{B}^2\bar{\mathbf{e}}^3 \wedge \bar{\mathbf{e}}^1 + \bar{B}^3\bar{\mathbf{e}}^1 \wedge \bar{\mathbf{e}}^2) \\
&= \left[\bar{E}^1\mathbf{e}^1 + \frac{\bar{E}^2 - v\bar{B}^3/c}{[1 - (v/c)^2]^{1/2}}\,\mathbf{e}^2 + \frac{\bar{E}^3 + v\bar{B}^2/c}{[1 - (v/c)^2]^{1/2}}\,\mathbf{e}^3\right] \wedge \mathbf{e}^4 \\
&\quad + \left[\bar{B}^1\mathbf{e}^2 \wedge \mathbf{e}^3 + \frac{\bar{B}^2 + v\bar{E}^3/c}{[1 - (v/c)^2]^{1/2}}\,\mathbf{e}^3 \wedge \mathbf{e}^1 + \frac{\bar{B}^3 - v\bar{E}^2/c}{[1 - (v/c)^2]^{1/2}}\,\mathbf{e}^1 \wedge \mathbf{e}^2\right],
\end{aligned} \tag{40.21}$$

we obtain

$$E^1 = \bar{E}^1, \qquad E^2 = \frac{\bar{E}^2 - v\bar{B}^3/c}{[1-(v/c)^2]^{1/2}}, \qquad E^3 = \frac{\bar{E}^3 + v\bar{B}^2/c}{[1-(v/c)^2]^{1/2}},$$
$$B^1 = \bar{B}^1, \qquad B^2 = \frac{\bar{B}^2 + v\bar{E}^3/c}{[1-(v/c)^2]^{1/2}}, \qquad B^3 = \frac{\bar{B}^3 - v\bar{E}^2/c}{[1-(v/c)^2]^{1/2}}. \tag{40.22}$$

Similarly, when we substitute (40.20) into the component form of $\mathbf{\Gamma}$ in the $(\bar{x}^i, c\bar{t})$ system, we get

$$H^1 = \bar{H}^1, \qquad H^2 = \frac{\bar{H}^2 + v\bar{D}^3/c}{[1-(v/c)^2]^{1/2}}, \qquad H^3 = \frac{\bar{H}^3 - v\bar{D}^2/c}{[1-(v/c)^2]^{1/2}},$$
$$D^1 = \bar{D}^1, \qquad D^2 = \frac{\bar{D}^2 - v\bar{H}^3/c}{[1-(v/c)^2]^{1/2}}, \qquad D^3 = \frac{\bar{D}^3 + v\bar{H}^2/c}{[1-(v/c)^2]^{1/2}}, \tag{40.23}$$

and when we substitute (40.20) into the component form of $\mathbf{\Psi}$ in the $(\bar{x}^i, c\bar{t})$ system, we obtain

$$j^1 = \frac{\bar{j}^1 - v\bar{q}}{[1-(v/c)^2]^{1/2}}, \qquad j^2 = \bar{j}^2, \qquad j^3 = \bar{j}^3, \qquad q = \frac{\bar{q} - v\bar{j}^1/c}{[1-(v/c)^2]^{1/2}}. \tag{40.24}$$

Clearly the transformation laws (40.22)–(40.24) are not the same as the classical transformation laws (32.32)–(32.34) and (32.39). Under the relativistic transformation laws Maxwell's equations are formally invariant; i.e., in the Lorentz system $(\bar{x}^i, c\bar{t})$ we still have

$$\overline{\text{curl}}\ \bar{\mathbf{E}} = -\frac{1}{c}\frac{\partial \bar{\mathbf{B}}}{\partial \bar{t}}, \qquad \overline{\text{div}}\ \bar{\mathbf{B}} = 0,$$
$$\overline{\text{curl}}\ \bar{\mathbf{H}} = 4\pi\frac{\bar{\mathbf{j}}}{c} + \frac{1}{c}\frac{\partial \bar{\mathbf{D}}}{\partial \bar{t}}, \qquad \overline{\text{div}}\ \bar{\mathbf{D}} = 4\pi\bar{q}. \tag{40.25}$$

Not only are Maxwell's equations formally invariant under any change of Lorentz system, so also are the Maxwell–Lorentz ether relations $\mathbf{D} = \mathbf{E}$ and $\mathbf{B} = \mathbf{H}$ for a vacuum. Indeed, the ether relations may be expressed directly in terms of the 2-forms $\mathbf{\Gamma}$ and $\mathbf{\Phi}$. Recall that for any oriented inner product space $\mathscr{U}$ we can define a duality operator $\mathbf{D}_r$, which maps a skew-symmetric tensor of order r to a skew-symmetric tensor of order $n - r$, where n denotes the dimensional of the underlying vector spaces $\mathscr{U}$. That operator is defined in Section 41, IVT-1. Now exactly the same definition may be applied to the oriented Minkowskian inner product space $\mathscr{V}$. As we shall see, the set of ether relations corresponds to the tensor equation

$$\mathbf{D}_2(\mathbf{\Phi}) = \mathbf{\Gamma}. \tag{40.26}$$

Specifically, for any integer r between 0 and 4, we define the duality operator $\mathbf{D}_r$ by the condition

$$\sigma_{4-r}(\mathbf{D}_r\mathbf{P}, \mathbf{Q}) = \sigma_4(\mathbf{P} \wedge \mathbf{Q}, \boldsymbol{\Xi}), \tag{40.27}$$

where $\mathbf{P}$ and $\mathbf{Q}$ are arbitrary skew-symmetric covariant tensors of orders r and $4 - r$, respectively. The symbol σ_k in (40.27) denotes the pseudo-inner-product of skew-symmetric covariant tensors of order k induced by σ; i.e., in component form

$$\sigma_k(\mathbf{P}, \mathbf{R}) \equiv \frac{1}{k!} \Sigma^{i_1 j_1} \cdots \Sigma^{i_k j_k} P_{i_1 \cdots i_k} R_{j_1 \cdots j_k}, \tag{40.28}$$

and the symbol $\boldsymbol{\Xi}$ denotes the positive density tensor on $\mathscr{V}$; i.e.,

$$\boldsymbol{\Xi} = \mathbf{e}^1 \wedge \mathbf{e}^2 \wedge \mathbf{e}^3 \wedge \mathbf{e}^4, \tag{40.29}$$

where $\{\mathbf{e}^\alpha\}$ is the dual basis of any Lorentz frame $\{\mathbf{e}_\alpha\}$. Notice that σ_1 is just the dual pseudo-inner-product σ^* defined by (37.5).

From (40.27)–(40.29) the duality operators $\mathbf{D}_1$, $\mathbf{D}_2$, and $\mathbf{D}_3$ are given explicitly as follows:

$$\begin{aligned} &\mathbf{D}_1(\mathbf{e}^1) = \mathbf{e}^2 \wedge \mathbf{e}^3 \wedge \mathbf{e}^4, \qquad &&\mathbf{D}_1(\mathbf{e}^2) = \mathbf{e}^3 \wedge \mathbf{e}^1 \wedge \mathbf{e}^4, \\ &\mathbf{D}_1(\mathbf{e}^3) = \mathbf{e}^1 \wedge \mathbf{e}^2 \wedge \mathbf{e}^4, &&\mathbf{D}_1(\mathbf{e}^4) = \mathbf{e}^1 \wedge \mathbf{e}^2 \wedge \mathbf{e}^3; \end{aligned} \tag{40.30}$$

$$\begin{aligned} &\mathbf{D}_2(\mathbf{e}^1 \wedge \mathbf{e}^2) = \mathbf{e}^3 \wedge \mathbf{e}^4, \qquad &&\mathbf{D}_2(\mathbf{e}^2 \wedge \mathbf{e}^3) = \mathbf{e}^1 \wedge \mathbf{e}^4, \\ &\mathbf{D}_2(\mathbf{e}^3 \wedge \mathbf{e}^1) = \mathbf{e}^2 \wedge \mathbf{e}^4, &&\mathbf{D}_2(\mathbf{e}^1 \wedge \mathbf{e}^4) = \mathbf{e}^3 \wedge \mathbf{e}^2, \\ &\mathbf{D}_2(\mathbf{e}^2 \wedge \mathbf{e}^4) = \mathbf{e}^1 \wedge \mathbf{e}^3, &&\mathbf{D}_2(\mathbf{e}^3 \wedge \mathbf{e}^4) = \mathbf{e}^2 \wedge \mathbf{e}^1; \end{aligned} \tag{40.31}$$

and

$$\begin{aligned} &\mathbf{D}_3(\mathbf{e}^1 \wedge \mathbf{e}^2 \wedge \mathbf{e}^3) = \mathbf{e}^4, \qquad &&\mathbf{D}_3(\mathbf{e}^1 \wedge \mathbf{e}^2 \wedge \mathbf{e}^4) = \mathbf{e}^3, \\ &\mathbf{D}_3(\mathbf{e}^2 \wedge \mathbf{e}^3 \wedge \mathbf{e}^4) = \mathbf{e}^1, &&\mathbf{D}_3(\mathbf{e}^3 \wedge \mathbf{e}^1 \wedge \mathbf{e}^4) = \mathbf{e}^2; \end{aligned} \tag{40.32}$$

where $\{\mathbf{e}^\alpha\}$ is the dual basis of any positive orthonormal basis $\{\mathbf{e}_\alpha\}$.

Note. A positive orthonormal basis $\{\mathbf{e}_\alpha\}$ is either a Lorentz frame or the negative of a Lorentz frame. The conditions (40.30)–(40.32) are clearly invariant when $\{\mathbf{e}_\alpha\}$ is replaced by $\{-\mathbf{e}_\alpha\}$.

Using (40.31) and the component forms (40.8) and (40.16), we see that the set of ether relations $\mathbf{D} = \mathbf{E}$ and $\mathbf{B} = \mathbf{H}$ corresponds precisely to the tensor equation (40.26), which is independent of any choice of Lorentz frame. As a result, the ether relations are invariant under any change of Lorentz system. This transformation property removes the difficulty mentioned before in the classical theory of electromagnetism.

In general a system of constitutive relations of the isotropic form

$$\mathbf{D} = \varepsilon \mathbf{E}, \qquad \mathbf{B} = \mu \mathbf{H} \tag{40.33}$$

is not formally invariant under a change of Lorentz system, however, unless the constants ε and μ satisfy the condition

$$\varepsilon\mu = 1. \tag{40.34}$$

When the condition (40.34) holds, the system (40.33) corresponds to the tensor equation

$$\varepsilon \mathbf{D}_2(\mathbf{\Phi}) = \mathbf{\Gamma}. \tag{40.35}$$

When the condition (40.34) does not hold, the constitutive relations in another Lorentz system $(\bar{x}^i, c\bar{t})$ may be derived from (40.33) by using the transformation laws for the fields **B**, **H**, **D**, and **E**.

To illustrate the transformation law of the constitutive relations, we write (40.33) in the component form

$$D^i = \varepsilon E^i, \qquad H^i = \frac{1}{\mu} B^i. \tag{40.36}$$

Then from (40.22) and (40.23) the relations (40.36) are transformed into

$$\bar{\mathbf{D}} = \boldsymbol{\varphi}\bar{\mathbf{E}} + \boldsymbol{\psi}\bar{\mathbf{B}}, \qquad \bar{\mathbf{H}} = \boldsymbol{\psi}\bar{\mathbf{E}} + \boldsymbol{\chi}\bar{\mathbf{B}}, \tag{40.37}$$

where $\boldsymbol{\varphi}$ and $\boldsymbol{\chi}$ are diagonal matrices:

$$\begin{aligned} \boldsymbol{\varphi} &= \operatorname{diag}\left(\varepsilon, \frac{\varepsilon - v^2/c^2\mu}{1-(v/c)^2}, \frac{\varepsilon - v^2/c^2\mu}{1-(v/c)^2}\right), \\ \boldsymbol{\chi} &= \operatorname{diag}\left(\frac{1}{\mu}, \frac{1/\mu - \varepsilon v^2/c^2}{1-(v/c)^2}, \frac{1/\mu - \varepsilon v^2/c^2}{1-(v/c)^2}\right), \end{aligned} \tag{40.38}$$

while $\boldsymbol{\psi}$ is a skew-symmetric matrix:

$$[\boldsymbol{\psi}] = \begin{bmatrix} 0 & 0 & 0 \\ 0 & 0 & \dfrac{(v/c)[(1/\mu) - \varepsilon]}{1-(v/c)^2} \\ 0 & -\dfrac{(v/c)[(1/\mu) - \varepsilon]}{1-(v/c)^2} & 0 \end{bmatrix}. \tag{40.39}$$

Clearly $\boldsymbol{\psi}$ vanishes, and $\boldsymbol{\varphi}$ and $\boldsymbol{\chi}$ reduce to $\varepsilon\mathbf{I}$ and $(1/\mu)\mathbf{I}$, respectively, when the condition (40.34) holds.

Note. The set of constitutive relations (40.36) corresponds to a tensor equation of the form

$$\mathbf{K}(\mathbf{\Phi}) = \mathbf{\Gamma}, \tag{40.40}$$

where $\mathbf{K}$ is a certain automorphism of the 6-dimensional space of skew-symmetric covariant tensors of order 2 over the 4-dimensional translation space $\mathscr{V}$ of $\mathscr{E}$. The tensor equation (40.40) has the component form (40.33) in all Lorentz systems if and only if $\mathbf{K}$ is proportional to $\mathbf{D}_2$ as shown in (40.35).

Since the Minkowskian inner product σ is invariant under any change of Lorentz system, the squared pseudonorm $\sigma_2(\mathbf{\Phi}, \mathbf{\Phi})$ of $\mathbf{\Phi}$ is always given by

$$\sigma_2(\mathbf{\Phi}, \mathbf{\Phi}) = \frac{1}{2!}\, \Sigma^{\alpha\beta}\Sigma^{\gamma\lambda}\Phi_{\alpha\gamma}\Phi_{\beta\lambda} = -\mathbf{E}\cdot\mathbf{E} + \mathbf{B}\cdot\mathbf{B}. \tag{40.41}$$

Hence

$$-\mathbf{E}\cdot\mathbf{E} + \mathbf{B}\cdot\mathbf{B} = -\bar{\mathbf{E}}\cdot\bar{\mathbf{E}} + \bar{\mathbf{B}}\cdot\bar{\mathbf{B}} \tag{40.42}$$

under any change of Lorentz frame. The identity may be proved directly by using the transformation laws (40.22) also. Similarly, from (40.31) and (40.41)

$$\begin{aligned} \sigma_2(\mathbf{\Phi}, \mathbf{D}_2(\mathbf{\Phi})) &= -2\mathbf{B}\cdot\mathbf{E} = -2\bar{\mathbf{B}}\cdot\bar{\mathbf{E}}, \\ \sigma_2(\mathbf{\Gamma}, \mathbf{\Gamma}) &= -\mathbf{H}\cdot\mathbf{H} + \mathbf{D}\cdot\mathbf{D} = -\bar{\mathbf{H}}\cdot\bar{\mathbf{H}} + \bar{\mathbf{D}}\cdot\bar{\mathbf{D}}, \\ \sigma_2(\mathbf{\Gamma}, \mathbf{D}_2(\mathbf{\Gamma})) &= 2\mathbf{D}\cdot\mathbf{H} = 2\bar{\mathbf{D}}\cdot\bar{\mathbf{H}}. \end{aligned} \tag{40.43}$$

Finally, the transformation laws of the constitutive relations must be consistent with the following identities:

$$\sigma_2(\mathbf{\Phi}, \mathbf{\Gamma}) = -\mathbf{E}\cdot\mathbf{D} - \mathbf{B}\cdot\mathbf{D} = -\bar{\mathbf{E}}\cdot\bar{\mathbf{H}} - \bar{\mathbf{B}}\cdot\bar{\mathbf{D}}, \tag{40.44a}$$

$$\sigma_2(\mathbf{\Phi}, \mathbf{D}_2(\mathbf{\Gamma})) = \mathbf{B}\cdot\mathbf{H} - \mathbf{E}\cdot\mathbf{D} = \bar{\mathbf{B}}\cdot\bar{\mathbf{H}} - \bar{\mathbf{E}}\cdot\bar{\mathbf{D}}. \tag{40.44b}$$

These identities may be proved for the special cases (40.33) and (40.37) directly from (40.38) and (40.39).

41. Lorentz's Formula and the Balance Principles in Special Relativistic Form

In the classical theory the force acting on a moving distribution of charge q in an electromagnetic field is given by Lorentz's formula

$$\mathbf{f} = q\left(\mathbf{E} + \frac{\mathbf{v}}{c} \times \mathbf{B}\right) = q\mathbf{E} + \frac{\mathbf{j}}{c} \times \mathbf{B}. \tag{41.1}$$

Then the power of the force is given by

$$p = \mathbf{f} \cdot \mathbf{v} = q\mathbf{v} \cdot \mathbf{E} = \mathbf{j} \cdot \mathbf{E}. \tag{41.2}$$

We shall now derive the relativistic versions of these formulas.

First, we show that there is a vector $\boldsymbol{\xi}$ which has the decomposition

$$\boldsymbol{\xi} = q\left[\mathbf{E} + \frac{\mathbf{v}}{c} \times \mathbf{B} + \left(\frac{\mathbf{v}}{c} \cdot \mathbf{E}\right)\mathbf{e}_4\right] = \left(q\mathbf{E} + \frac{\mathbf{j}}{c} \times \mathbf{B}\right) + \left(\frac{\mathbf{j}}{c} \cdot \mathbf{E}\right)\mathbf{e}_4 \tag{41.3}$$

relative to any Lorentz frame $\{\mathbf{e}_i, \mathbf{e}_4\}$. Notice that $\boldsymbol{\xi}$ satisfies the condition of orthogonality (39.27), viz.,

$$i\mathbf{v} \cdot q\left(\mathbf{E} + \frac{\mathbf{v}}{c} \times \mathbf{B}\right) - ciq\left(\frac{\mathbf{v}}{c} \cdot \mathbf{E}\right) = 0, \tag{41.4}$$

where we have used the formula (39.19).

To show that there is a vector $\boldsymbol{\xi}$ given by (41.3) relative to any choice of Lorentz frame $\{\mathbf{e}_i, \mathbf{e}_4\}$ we start from the relativistic charge–current field $\boldsymbol{\Psi}$ defined by the component form (40.15). Using the duality operator $\mathbf{D}_3$, we transform $\boldsymbol{\Psi}$ into a 1-form $\mathbf{J}$, viz.,

$$\mathbf{J} \equiv \mathbf{D}_3\boldsymbol{\Psi} = \frac{\mathbf{j}^1}{c}\,\mathbf{e}^1 + \frac{\mathbf{j}^2}{c}\,\mathbf{e}^2 + \frac{\mathbf{j}^3}{c}\,\mathbf{e}^3 - q\mathbf{e}^4. \tag{41.5}$$

Next, since a 2-form is just a skew-symmetric covariant tensor field of order 2, the electromagnetic field $\boldsymbol{\Phi}$ defined by (40.8) has the tensor component matrix

$$[\Phi_{\mu\nu}] = \begin{bmatrix} 0 & B^3 & -B^2 & E^1 \\ -B^3 & 0 & B^1 & E^2 \\ B^2 & -B^1 & 0 & E^3 \\ -E^1 & -E^2 & -E^3 & 0 \end{bmatrix}. \tag{41.6}$$

Using the operation of raising of indices $\boldsymbol{\Sigma}^*$, we transform this covariant tensor into a contravariant tensor with the component matrix

$$[\Phi^{\mu\nu}] = [(\Sigma^*\Phi\Sigma^*)^{\mu\nu}] = \begin{bmatrix} 0 & B^3 & -B^2 & -E^1 \\ -B^3 & 0 & B^1 & -E^2 \\ B^2 & -B^1 & 0 & -E^3 \\ E^1 & E^2 & E^3 & 0 \end{bmatrix}. \tag{41.7}$$

Now we apply the contravariant tensor field to the 1-form $\mathbf{D}_3\boldsymbol{\Psi}$, and we obtain a vector field

$$\begin{aligned} \boldsymbol{\Sigma}^*\boldsymbol{\Phi}\boldsymbol{\Sigma}^*\mathbf{D}_3\boldsymbol{\Psi} \\ &= \left(B^3\frac{j^2}{c} - B^2\frac{j^3}{c} + E^1 q\right)\mathbf{e}_1 + \left(B^1\frac{j^3}{c} - B^3\frac{j^1}{c} + E^2 q\right)\mathbf{e}_2 \\ &\quad + \left(B^2\frac{j^1}{c} - B^1\frac{j^2}{c} + E^3 q\right)\mathbf{e}_3 + \left(E^1\frac{j^1}{c} + E^2\frac{j^2}{c} + E^3\frac{j^3}{c}\right)\mathbf{e}_4 \\ &= \left(q\mathbf{E} + \frac{\mathbf{j}}{c}\times\mathbf{B}\right) + \left(\frac{\mathbf{j}}{c}\cdot\mathbf{E}\right)\mathbf{e}_4 = \boldsymbol{\xi}. \end{aligned} \tag{41.8}$$

Thus $\boldsymbol{\xi}$ is well defined by the decomposition (41.3); we call $\boldsymbol{\xi}$ the *momentum-energy production vector* of the electromagnetic field.

Next, we show that the momentum-energy production vector may be obtained from the divergence of a certain tensor field $\boldsymbol{\Omega}$, known as the *electromagnetic stress-energy-momentum tensor*. Specifically, we have the tensor equation

$$\operatorname{Div}\boldsymbol{\Omega} + \boldsymbol{\xi} = \mathbf{0}, \tag{41.9}$$

which corresponds to the system

$$\frac{\partial\Omega^{ij}}{\partial x^j} + \frac{1}{c}\frac{\partial\Omega^{i4}}{\partial t} = \xi^i, \qquad \frac{\partial\Omega^{4j}}{\partial x^j} + \frac{1}{c}\frac{\partial\Omega^{44}}{\partial t} = \xi^4. \tag{41.10}$$

As we shall see, these equations are just the balance equations (33.16) and (33.9) in the classical theory. As before we assume that the medium is characterized by a linear constitutive relation of the form (40.40), where $\mathbf{K}$ is a constant tensor. Under this assumption we define $\boldsymbol{\Omega}$ by the component matrix

$$[\boldsymbol{\Omega}] = \left[\begin{array}{c|c} \mathbf{T} & \dfrac{1}{4\pi}\mathbf{D}\times\mathbf{B} \\ \hline \mathbf{S}/c & \dfrac{1}{8\pi}(\mathbf{E}\cdot\mathbf{D} + \mathbf{H}\cdot\mathbf{B}) \end{array}\right], \tag{41.11}$$

where $\mathbf{T}$ is the Maxwell stress tensor [cf. (33.13)], $\mathbf{S}$ is the Poynting vector [cf. (33.8)], $(1/4\pi c)(\mathbf{D}\times\mathbf{B})$ is the electromagnetic momentum density [cf. (33.12)], and $(1/8\pi)(\mathbf{E}\cdot\mathbf{D}+\mathbf{H}\cdot\mathbf{B})$ is the electromagnetic energy density [cf. (33.7)].

We prove first that the right-hand side of (41.11) is, in fact, the component matrix of a contravariant tensor of order 2 on $\mathscr{V}$. Indeed, $\mathbf{\Omega}$ may be obtained from the tensors $\mathbf{\Sigma}^*$, $\mathbf{D}_2\mathbf{\Gamma}$, and $\mathbf{\Phi}$ in the following way: First, we define a tensor $\mathbf{W}$ by the component form

$$W^{\alpha\beta} = \Sigma^{\alpha\delta}\Phi_{\delta\mu}\Sigma^{\mu\lambda}(D_2\Gamma)_{\lambda\nu}\Sigma^{\nu\beta}. \tag{41.12}$$

From (40.16) and (40.31) the component matrix of the 2-form $\mathbf{D}_2\mathbf{\Gamma}$ is

$$[(D_2\Gamma)_{\lambda\nu}] = \begin{bmatrix} 0 & -H^3 & H^2 & -D^1 \\ H^3 & 0 & -H^1 & -D^2 \\ -H^2 & H^1 & 0 & -D^3 \\ D^1 & D^2 & D^3 & 0 \end{bmatrix}. \tag{41.13}$$

Substituting (41.7) and (41.13) into (41.12), we obtain

$$[W^{\alpha\beta}] = \begin{bmatrix} B^3H^3+B^2H^2-E^1D^1 & -B^2H^1-E^1D^2 & -B^3H^1-E^1D^3 & B^3D^2-B^2D^3 \\ -B^1H^2-E^2D^1 & B^3H^3+B^1H^1-E^2D^2 & -B^3H^2-E^2D^3 & B^3D^1-B^1D^3 \\ -B^1H^3-E^3D^1 & -B^2H^3-E^3D^2 & B^2H^2+B^1H^1-E^3D^3 & B^1D^2-B^2D^1 \\ E^2H^3-E^3H^2 & E^3H^1-E^1H^3 & E^1H^2-E^2H^1 & E^1D^1+E^2D^2+E^3D^3 \end{bmatrix}. \tag{41.14}$$

Now using the identity (40.44b) and the definitions (33.8) and (33.13), we see that

$$\Omega^{\mu\nu} = \frac{1}{4\pi}\left[W^{\mu\nu} + \frac{1}{2}\sigma_2(\mathbf{\Phi}, \mathbf{D}_2\mathbf{\Gamma})\Sigma^{\mu\nu}\right]. \tag{41.15}$$

Thus the tensor $\mathbf{\Omega}$ is well defined by (41.11).

Taking the divergence of $\mathbf{\Omega}$ from the component form (41.11), we get

$$\begin{aligned}\operatorname{Div}\mathbf{\Omega} &= \left[\operatorname{div}\mathbf{T} + \frac{1}{4\pi c}\frac{\partial}{\partial t}(\mathbf{D}\times\mathbf{B})\right] \\ &\quad + \frac{1}{c}\left[\operatorname{div}\mathbf{S} + \frac{1}{4\pi}\left(\mathbf{E}\cdot\frac{\partial\mathbf{D}}{\partial t} + \mathbf{H}\cdot\frac{\partial\mathbf{B}}{\partial t}\right)\right]\mathbf{e}_4.\end{aligned} \tag{41.16}$$

Then from the balance equations (33.9) and (33.16)

$$\operatorname{Div}\mathbf{\Omega} = -\left(q\mathbf{E} + \frac{\mathbf{j}}{c}\times\mathbf{B}\right) - \left(\frac{\mathbf{j}}{c}\cdot\mathbf{E}\right)\mathbf{e}_4 = -\boldsymbol{\xi}. \tag{41.17}$$

Thus the tensor equation (41.9) is proved.

Note. The classical balance equations (33.9) and (33.16) may be used here, because they are actually mathematical identities for all solutions of Maxwell's equations. Since the field equations relative to a Lorentz frame in the relativistic theory are exactly the same as the field equations relative to the ether frame in the classical theory, those identities remain valid in special relativity.

As remarked in Section 33, if we use Lorentz's electron theory, then the constitutive relations are always the same as the ether relations, provided that the charge–current field includes the bound charge and the magnetization current in the medium. In application that theory is not convenient, since the bound charge and the magnetization current are not known *a priori*, and we still have to use the conventional constitutive relations to determine them. Conceptually, however, the electron theory is easier to formulate, since the bound charge and the magnetization current are governed by the same laws of physics as those for the free charge and the free current. In other words the electron theory gives the "true" electromagnetic fields in the interior of a medium (i.e., at the empty space not occupied by the electrons). In the general theory of relativity, which we shall consider in the next two chapters, it is the "true" electromagnetic stress-energy-momentum tensor based on the electron theory, not the conventional stress-energy-momentum tensor $\mathbf{\Omega}$ defined by (41.11), that gives rise to an electromagnetic–gravitational interaction.

When we use the ether relations as the constitutive relations, the balance equation (41.17) is replaced by

$$\operatorname{Div} \hat{\mathbf{\Omega}} = -\boldsymbol{\xi} = -\left(\hat{q}\mathbf{E} + \frac{\hat{\mathbf{j}}}{c} \times \mathbf{B}\right) - \left(\frac{\hat{\mathbf{j}}}{c} \cdot \mathbf{E}\right)\mathbf{e}_4, \tag{41.18}$$

where $\hat{\mathbf{j}}$ and $\hat{\mathbf{q}}$ are given as before by (33.18), and where $\hat{\mathbf{\Omega}}$ is a symmetric tensor having the component form

$$[\hat{\mathbf{\Omega}}] = \left[\begin{array}{c|c} \hat{\mathbf{T}} & \dfrac{1}{4\pi}\,\mathbf{E}\times\mathbf{B} \\ \hline \dfrac{1}{4\pi}\,\mathbf{E}\times\mathbf{B} & \dfrac{1}{8\pi}\,(E^2 + B^2) \end{array}\right], \tag{41.19}$$

where $\hat{\mathbf{T}}$ is defined by (33.22). The fact that the right-hand side of (41.19) is the component matrix of a contravariant tensor of order 2 on $\mathscr{V}$ may be verified easily as before.

42. Doppler Effect of Electromagnetic Waves

As remarked in the preceding two sections, the field equations of electromagnetism relative to a Lorentz frame are exactly the same as those in the classical theory relative to the ether frame. In particular, all classical results associated with the ether frame remain valid in the special theory of relativity, provided that they are interpreted as results associated with a Lorentz frame. For instance the result that the speed of light in a vacuum is the constant c in the ether frame becomes the result that the speed of light in a vacuum is the constant c in any Lorentz frame.

Most classical results pertaining to a moving frame fail to remain valid in the relativistic theory, however. In the classical theory the field equations relative to a moving frame are obtained from those of the ether frame by certain transformation laws, which are established by direct argument based on the nature of the field quantities involved. Unfortunately, the constitutive relations, such as the ether relations, are not invariant under a change of frame by a Galilean transformation. As a result, according to the classical theory the speed of light in vacuum is not equal to the constant c in a moving frame. From the famous experiment of Michelson and Morley this prediction is known to be inconsistent with the observation. This difficulty is removed by the theory of special relativity, since the field equations are the same in all Lorentz frames. In fact the transformation laws of the field quantities are derived from that condition, and they are not the same as those in the classical theory.

To illustrate the relativistic transformation laws, we consider the Doppler effect for electromagnetic waves. For simplicity we use the special Lorentz transformation given by (38.2). Setting $x^1 = x^2 = x^3 = 0$, we see that

$$\bar{x}^1 = v\bar{t}, \qquad \bar{x}^2 = \bar{x}^3 = 0. \tag{42.1}$$

Thus the unbarred frame is moving with speed v along the $\bar{x}^1$ axis in the barred frame. We consider a sinusoidal electromagnetic wave emitting from the origin of the unbarred frame in the direction

$$\mathbf{n} = -\mathbf{e}_1. \tag{42.2}$$

We shall now explain the behavior of this wave as observed in the barred system $(\bar{x}^i, c\bar{t})$.

In Section 34 of the preceding chapter we have shown that the electromagnetic fields $\mathbf{E}$ and $\mathbf{H}$ associated with a polarized, monochromatic, sinusoidal wave in the direction $-\mathbf{e}_1$ have the following component forms:

$$\mathbf{E} = \alpha \sin[k(x^1 + ct)]\mathbf{e}_3, \qquad \mathbf{H} = \alpha \sin[k(x^1 + ct)]\mathbf{e}_2, \tag{42.3}$$

where k denotes the wave number and where α denotes the amplitude. The fields $\mathbf{E}$ and $\mathbf{H}$ given by (42.3) satisfy Maxwell's equations in vacuum.

Now to determine the fields of the same wave as observed in the barred system $(\bar{x}^i, c\bar{t})$, we use the inverse of the transformation laws (40.22) and (40.23). Specifically, we have

$$\begin{aligned}\bar{E}^3 = \bar{H}^2 &= \frac{E^3 - vB^2/c}{[1 - (v/c)^2]^{1/2}} \\ &= \left(\frac{1 - v/c}{1 + v/c}\right)^{1/2} \alpha \sin\left\{k\, \frac{\bar{x}^1 - v\bar{t} + c\bar{t} - v\bar{x}^1/c}{[1 - (v/c)^2]^{1/2}}\right\} \\ &= \left(\frac{1 - v/c}{1 + v/c}\right)^{1/2} \alpha \sin\left[k\left(\frac{1 - v/c}{1 + v/c}\right)^{1/2} (\bar{x}^1 + c\bar{t})\right] \\ &= \bar{\alpha} \sin \bar{k}(\bar{x}^1 + c\bar{t}),\end{aligned} \tag{42.4}$$

where

$$\bar{\alpha} = \alpha\left(\frac{1 - v/c}{1 + v/c}\right)^{1/2}, \qquad \bar{k} = k\left(\frac{1 - v/c}{1 + v/c}\right)^{1/2}. \tag{42.5}$$

The components $\bar{E}^i$ and $\bar{H}^i$ not shown in (42.4) all vanish, viz.,

$$\bar{\mathbf{E}} = \bar{E}^3\bar{\mathbf{e}}_3, \qquad \bar{\mathbf{H}} = \bar{H}^2\bar{\mathbf{e}}_2. \tag{42.6}$$

Notice that the component forms (42.4) imply that in the barred system $(\bar{x}^i, c\bar{t})$ the electromagnetic fields still correspond to a sinusoidal wave, which is propagating in the direction $\bar{\mathbf{n}} = -\bar{\mathbf{e}}_1$, and the wave speed is c. We can verify that the identities (40.42) and (40.43) are satisfied. In fact in this case

$$\bar{\mathbf{E}} \cdot \bar{\mathbf{H}} = \mathbf{E} \cdot \mathbf{H} = 0, \qquad \bar{E}^2 - \bar{H}^2 = E^2 - H^2 = 0. \tag{42.7}$$

For any sinusoidal wave the *Doppler effect* is the fact that the frequency of the wave depends on the motion of the observer relative to the wave. Let ω and $\bar{\omega}$ denote the frequencies of the wave as observed in the unbarred frame and in the barred frame, respectively. Then from (34.8) and (42.5) we see that $\bar{\omega}$ differs from ω by a factor $[(1 + v/c)/(1 - v/c)]^{1/2}$. When $v > 0$, i.e., when the light source is moving away from the observer, the factor $[(1 + v/c)/(1 - v/c)]^{1/2}$ has a value greater than 1. Thus $\bar{\omega} > \omega$. This situation is known as a *red shift*. We can use the value of the ratio $\bar{\omega}/\omega$ to calculate the speed v of the light source as it moves away from the observer. When $v < 0$, i.e., when the light source is moving towards the observer, the factor $[(1 + v/c)/(1 - v/c)]^{1/2}$ has a value less than 1. Thus $\bar{\omega} < \omega$. This situation is known as a *blue shift*.

Of course, the classical theory predicts a Doppler effect for an electromagnetic wave also. But the classical transformation laws imply that

$$\bar{\alpha} = \alpha, \qquad \bar{k} = k\,\frac{1}{1 + v/c}, \tag{42.8}$$

and that the wave speed observed in the unbarred frame differs from that in the barred frame by the relative speed v of the two frames. These results are clearly not the same as those predicted by the relativistic theory.

When the light source is moving in a direction perpendicular to the direction of the light beam, the classical theory predicts no Doppler effect at all. Such is not the case if the theory of relativity is used. Consider the special case when the wave is propagating in the direction $\mathbf{n} = \mathbf{e}_3$ and is polarized in such a way that $\mathbf{E}$ is parallel to $\mathbf{e}_2$ while $\mathbf{H}$ is parallel to $\mathbf{e}_1$, viz.,

$$\mathbf{E} = \alpha \sin[k(x^3 - ct)]\mathbf{e}_2, \qquad \mathbf{H} = \alpha \sin[k(x^3 - ct)]\mathbf{e}_1. \tag{42.9}$$

For this wave the inverse of the transformation laws (40.22) and (40.23) implies that

$$\bar{\mathbf{E}} = \bar{\alpha} \sin[\bar{k}(\bar{\mathbf{n}} \cdot \bar{\mathbf{x}} - c\bar{t})], \tag{42.10}$$

where

$$\begin{aligned} \bar{\alpha} &= \frac{\alpha}{[1 - (v/c)^2]^{1/2}}, \\ \bar{k} &= \frac{k}{[1 - (v/c)^2]^{1/2}}, \\ \bar{\mathbf{n}} &= (v/c)\bar{\mathbf{e}}_1 + [1 - (v/c)^2]^{1/2}\bar{\mathbf{e}}_3. \end{aligned} \tag{42.11}$$

Then $\bar{\mathbf{H}}$ may be obtained directly by

$$\bar{\mathbf{H}} = \bar{\mathbf{n}} \times \bar{\mathbf{E}}. \tag{42.12}$$

A similar problem with $\mathbf{E}$ parallel to $\mathbf{e}_1$ and $\mathbf{H}$ parallel to $-\mathbf{e}_2$ may be formulated and solved in the same way. Again, there is a factor $1/[1 - (v/c)^2]^{1/2}$ connecting $\bar{\alpha}$, $\bar{k}$ with α, k as shown in (42.11). This factor is of second order in (v/c) and is strictly a relativistic effect.

7

General Relativistic Theory of Gravitation

The special theory of relativity summarized in the preceding chapter was formulated by Einstein in order to resolve the difficulties in the classical theory of electromagnetism with regard to a change of inertial frame and the associated change of the speed of light. The special status of the inertial frames was not removed, however, but was transferred to the Lorentz frames. In the general theory of relativity that special status was finally removed, and a relativistic theory of gravitation was created. We shall summarize the mathematical structure of the general theory and its major results in this chapter.

43. Newton's Law of Gravitation and the Principle of Equivalence

In classical mechanics the magnitude of the gravitational attraction between two particles is given by *Newton's law*:

$$f = k \frac{m_1 m_2}{r^2}, \tag{43.1}$$

where m_1 and m_2 denote the masses of the two particles, and where r denotes the distance separating the particles. The empirical value of the constant k is

$$k = 6.66 \times 10^{-8}\ \text{cm}^3\ \text{g}^{-1}\ \text{sec}^{-2} \tag{43.2}$$

in cgs units. The line of the forces of attraction passes through the locations of the two particles.

Since f is proportional to m_1 and m_2, it may be characterized by a gravitational field. Specifically, let $\mathbf{g}$ be a vector field such that if we introduce a particle with mass $\bar{m}$ into its domain, then the gravitational force acting on that particle is $\bar{m}\mathbf{g}$. We call $\mathbf{g}$ the *gravitational field.* From Newton's law (43.1) the gravitational field $\mathbf{g}$ due to a particle with mass m is

$$\mathbf{g} = \mathbf{g}(\mathbf{x}) = -k \frac{m}{r^3} \mathbf{r}, \tag{43.3}$$

where $\mathbf{r}$ denotes the position vector of the field point $\mathbf{x}$ relative to the location of the particle. The minus sign in the formula (43.3) means that particles attract one another by gravitation.

It follows from (43.3) that $\mathbf{g}$ is a conservative field which may be expressed as

$$\mathbf{g} = -\text{grad}\ \zeta, \tag{43.4}$$

where the gravitational potential ζ is given by

$$\zeta = \zeta(\mathbf{x}) = -\frac{km}{r}. \tag{43.5}$$

The preceding formulation shows that a particle plays two distinct roles relative to a gravitational field. First, a particle is a source point for the gravitational field; i.e., the presence of a particle gives rise to a gravitational field. Second, a particle is also a receptor point for the gravitational field; i.e., the presence of a gravitational field gives rise to a force on a particle as soon as that particle is introduced into the domain of the field.

The formulation of the gravitational field associated with a particle may be generalized to that associated with a distribution of mass with density ϱ on some domain $\mathscr{D}_\varrho$. The previous formula (43.3) is generalized to

$$\mathbf{g}(\mathbf{x}) = \int_{\mathscr{D}_\varrho} -\frac{k\varrho}{r^3}\mathbf{r}\, dv = -\text{grad}\left(-\int_{\mathscr{D}_\varrho} \frac{k\varrho}{r}\, dv\right), \tag{43.6}$$

where dv denotes the Euclidean volume element at the position $\mathbf{x} - \mathbf{r}$ in $\mathscr{D}_\varrho$. The formula (43.6) shows that $\mathbf{g}$ is again a conservative field, and that the gravitational potential is given by

$$\zeta(\mathbf{x}) = -\int_{\mathscr{D}_\varrho} \frac{k\varrho}{r}\, dv. \tag{43.7}$$

In particular, $\mathbf{g}$ satisfies the field equations

$$\operatorname{curl} \mathbf{g} = \mathbf{0}, \tag{43.8a}$$

$$\operatorname{div} \mathbf{g} = -4\pi k \varrho, \tag{43.8b}$$

where we have used the formula (30.8).

Substituting (43.6) into (43.8), we see that ζ satisfies the Poisson equation

$$\nabla^2 \zeta = 4\pi k \varrho. \tag{43.9}$$

As usual we require that $\zeta \to 0$ at infinity. Then the solution of the preceding field equation is given by (43.7). We can integrate the field equation (43.8b) over any domain $\mathscr{D}$ in space, obtaining

$$\int_{\mathscr{D}} \operatorname{div} \mathbf{g}\, dx = \int_{\partial\mathscr{D}} \mathbf{g} \cdot \mathbf{n}\, d\sigma = -4\pi k \int_{\mathscr{D}} \varrho\, dx, \tag{43.10}$$

where dx denotes the Euclidean volume element at the field point $\mathbf{x}$. We call the surface integral in (43.10) the *gravitational flux* through $\partial\mathscr{D}$. Then the formula (43.10) shows that this flux determines the amount of mass enclosed by the surface $\partial\mathscr{D}$. This result is similar to the Gauss flux theorem in electrostatics.

In the preceding formulation it is understood that the gravitational field $\mathbf{g}$ is present simultaneously with the presence of mass. Thus the relations (43.3) and (43.6) are valid in the instantaneous spaces of the Newtonian space–time and are independent of any frame of reference. In order to describe the gravitational field $\mathbf{g}$, we may introduce a frame of reference, of course. Then the position of the particle or the mass distribution in the instantaneous space may be characterized as usual by the coordinates (x^i, t), and the gravitational field $\mathbf{g}$ may be represented by a vector field on the physical space $\mathscr{S}$ of the frame of reference. If the mass is at rest, then the corresponding gravitational field $\mathbf{g}$ is a steady field depending only on the spatial coordinates x^i. If the mass is moving relative to the frame, then $\mathbf{g}$ depends on both x^i and t in such a way that the relations (43.3) and (43.6) hold at each instant t.

Now suppose that a gravitational field $\mathbf{g}$ is present in $\mathscr{S}$, and let a particle with mass $\bar{m}$ be introduced into the domain of $\mathbf{g}$. Then a gravitational force $\bar{m}\mathbf{g}$ acts immediately on the particle. If this force is not balanced by a suitable support force, then a motion of the particle results. Indeed, if the frame of reference is an inertial one, then according to Newton's law of motion (4.5) the acceleration $\mathbf{a}$ of the particle is just the vector $\mathbf{g}$, viz.,

$$\mathbf{a} = \frac{d^2\mathbf{x}}{dt^2} = \frac{\bar{m}\mathbf{g}}{\bar{m}} = \mathbf{g}. \tag{43.11}$$

By virtue of this equation $\mathbf{g}$ is also called the *gravitational acceleration.*

As remarked in analytical mechanics, the equation of motion (43.11) is valid relative to an inertial frame of reference only. Suppose that the frame is not an inertial one, say, if the frame is moving with a translational acceleration $\mathbf{h}$ relative to an inertial frame. Then the equation of motion of the particle (relative to the noninertial frame) becomes

$$\bar{\mathbf{a}} + \mathbf{h} = \frac{\bar{m}\mathbf{g}}{\bar{m}} = \mathbf{g}, \tag{43.12}$$

or, equivalently,

$$\bar{\mathbf{a}} = \mathbf{g} - \mathbf{h}, \tag{43.13}$$

where $\bar{\mathbf{a}}$ denotes the acceleration of the particle as observed in the non-inertial frame. Comparing the preceding equation with the previous equation (43.11), we see that the gravitational acceleration $\mathbf{g}$ may be compensated by an inertial acceleration $\mathbf{h}$ due to the motion of the frame relative to an inertial frame. Physically this fact may be illustrated most vividly by using the concept of the "Einstein elevator." Indeed, an observer in an elevator which is falling freely in space cannot detect any acceleration $\bar{\mathbf{a}}$ of a particle inside the elevator, since $\mathbf{h} = \mathbf{g}$ in this case.

Since the physical interpretation of an inertial frame of reference is not really known in classical mechanics, when we observe an acceleration $\bar{\mathbf{a}}$ of a particle relative to a convenient frame of reference (e.g., the frame attached to the earth or the frame attached to the sun) we do not really know which part of the observed acceleration $\bar{\mathbf{a}}$ is due to a gravitational field and which part is due to the departure of the frame of reference from an inertial frame. In other words, the presence of a gravitational field $\mathbf{g}$ in a small domain in the physical space $\mathscr{S}$ of a frame of reference (e.g., inside the "Einstein elevator") is mathematically equivalent to the departure of the frame from an inertial frame. Relative to the frame attached to the "Einstein elevator" it makes no difference mathematically whether we interpret the equation

$$\bar{\mathbf{a}} = \mathbf{g} - \mathbf{h} = \mathbf{0} \tag{43.14}$$

by asserting that a gravitational field $\mathbf{g}$ is present inside the elevator but that it is compensated by the inertial acceleration $\mathbf{h}$ due to the motion of the frame, or by asserting that there is no gravitational field inside the elevator and that the frame is an inertial frame. This mathematical equivalence of the model of gravitation and the model of the departure of a frame of reference from an inertial frame is called the *principle of equivalence.*

It should be noted that the inertial acceleration $\mathbf{h}$ associated with the noninertial frame under consideration is a vector field on $\mathscr{S}$ depending only on t and not on x^i. Since $\mathbf{g}$ generally depends on both t and x^i, the equivalence of $\mathbf{g}$ with $\mathbf{h}$ is valid at a point (x^i, t) and is approximately valid in a small domain (e.g., the interior of the "Einstein elevator" in the gravitational field of the earth). For any one gravitational field $\mathbf{g}$ the "Einstein elevators" at different points (x^i, t) correspond to not just one (noninertial) frame of reference but to a field of (noninertial) frames of reference. This remark is one of the reasons which motivate the use of a *Minkowskian manifold* as a mathematical model for the event world in the general theory of relativity.

We define a *Minkowskian manifold* $\mathscr{E}$ as follows: a 4-dimensional differentiable manifold such that the tangent space $\mathscr{E}_{\mathbf{w}}$ at each point $\mathbf{w}$ in $\mathscr{E}$ is equipped with a Minkowskian inner product $\sigma(\mathbf{w})$ together with a Lorentzian orientation which identifies an equivalence class of orthonormal bases as the Lorentz bases at $\mathbf{w}$. It is no longer required that $\mathscr{E}$ be an affine space. Also, there may or may not be any coordinate system (x^α) in which the component matrix $[\Sigma_{\mu\nu}]$ of the Minkowskian inner product is the constant matrix diag$(1, 1, 1, -1)$. We require only that σ be a smooth field on $\mathscr{E}$.

If there is a coordinate system (x^α) on some domain $\mathscr{D}$ in $\mathscr{E}$ such that the component matrix $[\Sigma_{\mu\nu}]$ of σ is the constant matrix diag$(1, 1, 1, -1)$, then $\mathscr{D}$ is regarded as being free of gravitational field, and (x^α) is called a *Lorentz system* on $\mathscr{D}$. Such a coordinate system is similar to a (global) Lorentz system for the Minkowskian space–time, which may be regarded as a special case of a Minkowskian manifold that is free of gravitational field everywhere.

As explained in the preceding chapter, the Lorentz systems in special relativity are the counterparts of the inertial systems in classical mechanics. In general relativity the presence of an inertial system (i.e., a Lorentz system) is directly related to the absence of a gravitational field. This important feature is motivated by the principle of equivalence.

We shall summarize the mathematical structure of a Minkowskian manifold in general in the following section. On the basis of that structure we formulate Einstein's field equations in subsequent sections.

44. Minkowskian Manifold

As mentioned in the preceding section, a Minkowskian manifold is a 4-dimensional differentiable manifold $\mathscr{E}$ such that the tangent space $\mathscr{E}_{\mathbf{w}}$ at each point $\mathbf{w} \in \mathscr{E}$ is equipped with a particular Minkowskian inner product $\sigma(\mathbf{w})$ together with a Lorentzian orientation which designates a certain equivalence class of orthonormal bases as the Lorentz frames at the point $\mathbf{w}$. We now discuss such a structure in detail.

First, the differentiable structure on a 4-dimensional manifold $\mathscr{E}$ is defined by a maximal collection Δ of (local) coordinate systems (x^α, $\alpha = 1, 2, 3, 4$) such that the coordinate transformations are smooth. Specifically, any two systems in Δ, say, (x^α) and $(\bar{x}^\alpha)$, are related to each other by smooth functions

$$x^\alpha = x^\alpha(\bar{x}^\beta), \qquad \bar{x}^\alpha = \bar{x}^\alpha(x^\beta) \tag{44.1}$$

on the overlap of their domains, and Δ is maximal with respect to such transformations. We may define a differentiable structure formally by a pair $(\mathscr{E}, \Delta)$.

Using the differentiable structure, we define a tangent space $\mathscr{E}_{\mathbf{w}}$ at each point $\mathbf{w} \in \mathscr{E}$ as follows: We start from a coordinate system (x^α) on a coordinate neighborhood $\mathscr{D}$ of $\mathbf{w}$. Since the coordinate system is a diffeomorphism of $\mathscr{D}$ into $\mathscr{R}^4$, it gives rise to a one-to-one correspondence between smooth curves in $\mathscr{D}$ and smooth curves in the coordinate set of $\mathscr{D}$ in $\mathscr{R}^4$. We use this one-to-one correspondence to define an equivalence relation among smooth curves passing through the point $\mathbf{w}$. Specifically, we say that $\boldsymbol{\lambda}$ and $\boldsymbol{\mu}$ are equivalent (or have the same tangent vector at $\mathbf{w}$) if their coordinate forms have the same tangent vector at the coordinates $(x^\alpha(\mathbf{w}))$ of $\mathbf{w}$. By virtue of the smoothness requirement of the coordinate transformations, it can be verified easily that this equivalence relation is well defined; i.e., if the coordinate forms of $\boldsymbol{\lambda}$ and $\boldsymbol{\mu}$ have the same tangent vector relative to any one coordinate system $(x^\alpha) \in \Delta$, then they have the same tangent vector relative to all coordinate systems in Δ. Now we define the *tangent space* $\mathscr{E}_{\mathbf{w}}$ simply as the space formed by the equivalence classes defined by the preceding equivalence relation.

It is clear that the tangent space $\mathscr{E}_{\mathbf{w}}$ just defined has the structure of a 4-dimensional vector space. Indeed, there is an one-to-one correspondence between a tangent vector at $\mathbf{w}$ (i.e., an equivalence class of curves passing through $\mathbf{w}$) and a tangent vector at $(x^\alpha(\mathbf{w}))$ in $\mathscr{R}^4$ by the very definition of the tangent space. Hence we can define a structure of vector space on $\mathscr{E}_{\mathbf{w}}$ by requiring that this one-to-one correspondence be a (linear) isomorphism.

Again, the requirement that the coordinate transformations are smooth ensures us that the structure of vector space defined in this way is independent of the choice of the coordinate system.

Note. If we can regard the manifold $\mathscr{E}$ as a smooth 4-dimensional surface in a Euclidean space of some higher dimension, then the preceding formal definition of the tangent space $\mathscr{E}_{\mathbf{w}}$ is consistent with the elementary definition of the tangent space of a surface. In fact, it is known that every differentiable manifold can always be regarded as a smooth surface in a Euclidean space of sufficiently high dimension. We choose not to define the tangent space on the basis of this result, however, since the proof of the result is not elementary.

The coordinate curves of any coordinate system (x^α) in Δ are smooth curves, of course. We can verify that their tangent vectors $\{\mathbf{h}_\alpha(\mathbf{w})\}$ at the point $\mathbf{w}$ form a basis for the tangent space $\mathscr{E}_{\mathbf{w}}$. This result may be seen most easily by using the isomorphism of $\mathscr{E}_{\mathbf{w}}$ with $\mathscr{R}^4$ mentioned in the definition of the vector space structure on $\mathscr{E}_{\mathbf{w}}$. Indeed, the image of $\{\mathbf{h}_\alpha(\mathbf{w})\}$ under the isomorphism is just the standard basis of $\mathscr{R}^4$. As before we call the basis $\{\mathbf{h}_\alpha(\mathbf{w})\}$ the *natural basis* of (x^α) at $\mathbf{w}$ for the tangent space $\mathscr{E}_{\mathbf{w}}$.

The dual space $\mathscr{E}_{\mathbf{w}}^*$ of $\mathscr{E}_{\mathbf{w}}$ is called the *cotangent space* of $\mathscr{E}$ at $\mathbf{w}$. That space is spanned by the natural basis $\{\mathbf{h}^\alpha(\mathbf{w})\}$, which is formed by the differentials of the coordinate functions x^α at $\mathbf{w}$ and is also the dual basis of the basis $\{\mathbf{h}_\alpha(\mathbf{w})\}$ in $\mathscr{E}_{\mathbf{w}}$. The *differential* of a smooth function on a neighborhood $\mathscr{D} \ni \mathbf{w}$ may be defined in terms of the usual differential of a smooth function on $\mathscr{R}^4$ by using a coordinate system (x^α) in $\mathscr{D}$ as before. Specifically, we say that two functions f and g have the same differential at $\mathbf{w}$ if their coordinate representations $f(x^\alpha)$ and $g(x^\alpha)$ have the same differential at the coordinates $(x^\alpha(\mathbf{w}))$ of $\mathbf{w}$. Again, the smoothness of the coordinate transformations ensures us that the definition does not depend on the choice of coordinate system in Δ. Then $\mathscr{E}_{\mathbf{w}}^*$ is just the space of all differentials of smooth functions at the point $\mathbf{w}$.

Now a *Minkowskian metric* may be defined as follows: a smooth field σ whose value $\sigma(\mathbf{w})$ at each point $\mathbf{w} \in \mathscr{E}$ is a Minkowskian inner product on $\mathscr{E}_{\mathbf{w}}$. As before $\sigma(\mathbf{w})$ gives rise to the operations of lowering and raising of indices, $\mathbf{\Sigma}(\mathbf{w})\colon \mathscr{E}_{\mathbf{w}} \to \mathscr{E}_{\mathbf{w}}^*$ and $\mathbf{\Sigma}^*(\mathbf{w})\colon \mathscr{E}_{\mathbf{w}}^* \to \mathscr{E}_{\mathbf{w}}$, respectively. In component form relative to any coordinate system (x^α)

$$\mathbf{\Sigma} = \Sigma_{\mu\nu}\mathbf{h}^\mu \otimes \mathbf{h}^\nu, \qquad \mathbf{\Sigma}^* = \Sigma^{\mu\nu}\mathbf{h}_\mu \otimes \mathbf{h}_\nu, \tag{44.2}$$

where $\Sigma_{\mu\nu}$ and $\Sigma^{\mu\nu}$ are smooth functions of (x^α).

The fact that $\sigma(\mathbf{w})$ is a Minkowskian inner product on $\mathscr{E}_{\mathbf{w}}$ means that there is an orthonormal basis, say, $\{\mathbf{e}_\alpha(\mathbf{w})\}$, such that $\mathbf{\Sigma}(\mathbf{w})$ and $\mathbf{\Sigma}^*(\mathbf{w})$ have the component forms

$$\begin{aligned}\mathbf{\Sigma}(\mathbf{w}) &= \delta_{ij}\mathbf{e}^i(\mathbf{w}) \otimes \mathbf{e}^j(\mathbf{w}) - \mathbf{e}^4(\mathbf{w}) \otimes \mathbf{e}^4(\mathbf{w}),\\ \mathbf{\Sigma}^*(\mathbf{w}) &= \delta^{ij}\mathbf{e}_i(\mathbf{w}) \otimes \mathbf{e}_j(\mathbf{w}) - \mathbf{e}_4(\mathbf{w}) \otimes \mathbf{e}_4(\mathbf{w}).\end{aligned} \tag{44.3}$$

It should be noted, however, that there need not be a local coordinate system (x^α) such that the corresponding field of natural basis $\{\mathbf{h}_\alpha\}$ is a field of orthonormal bases. In particular, we can no longer take the component fields $\Sigma_{\mu\nu}$ and $\Sigma^{\mu\nu}$ to be the constant fields given by (39.7).

In Section 59, IVT-2, we showed that a necessary and sufficient condition for a Riemannian manifold to be locally Euclidean is that the curvature tensor based on the metric vanishes. Exactly the same argument shows that the curvature tensor based on the Minkowskian metric vanishes if and only if local Lorentz systems exist. To define the curvature tensor $\mathbf{R}$, we recall that the covariant derivative of a vector field $\mathbf{u}$ is given by the component formula

$$u^\alpha{}_{,\beta} = \frac{\partial u^\alpha}{\partial x^\beta} + \left\{\begin{matrix}\alpha\\ \gamma\beta\end{matrix}\right\} u^\gamma; \tag{44.4}$$

cf. equation (59.9) in Section 59, IVT-2. Then the Ricci identities for the second covariant derivative are

$$u^\alpha{}_{,\beta\gamma} - u^\alpha{}_{,\gamma\beta} = -u^\delta R^\alpha{}_{\delta\beta\gamma}; \tag{44.5}$$

cf. equation (59.9) in Section 59, IVT-2. The quantities $R^\alpha{}_{\delta\beta\gamma}$ are the components of the curvature tensor field $\mathbf{R}$ and are given explicitly by

$$R^\alpha{}_{\delta\beta\gamma} = \frac{\partial}{\partial x^\beta}\left\{\begin{matrix}\alpha\\ \delta\gamma\end{matrix}\right\} - \frac{\partial}{\partial x^\gamma}\left\{\begin{matrix}\alpha\\ \delta\beta\end{matrix}\right\} + \left\{\begin{matrix}\lambda\\ \delta\gamma\end{matrix}\right\}\left\{\begin{matrix}\alpha\\ \lambda\beta\end{matrix}\right\} - \left\{\begin{matrix}\lambda\\ \delta\beta\end{matrix}\right\}\left\{\begin{matrix}\alpha\\ \lambda\gamma\end{matrix}\right\}; \tag{44.6}$$

cf. equation (59.10) in Section 59, IVT-2.

The curvature tensor $\mathbf{R}$ satisfies the following symmetry conditions:

$$R^\alpha{}_{\delta\beta\gamma} + R^\alpha{}_{\beta\gamma\delta} + R^\alpha{}_{\gamma\delta\beta} = 0, \tag{44.7}$$

and

$$R_{\alpha\delta\beta\gamma} = -R_{\delta\alpha\beta\gamma} = -R_{\alpha\delta\gamma\beta} = R_{\beta\gamma\alpha\delta}, \tag{44.8}$$

where the covariant components $R_{\alpha\delta\beta\gamma}$ are obtained from the mixed components $R^\alpha{}_{\delta\beta\gamma}$ by using the operation of lowering of indices, viz.,

$$R_{\alpha\delta\beta\gamma} = \Sigma_{\alpha\theta} R^\theta{}_{\delta\beta\gamma}. \tag{44.9}$$

The conditions (44.8) mean that $\mathbf{R}(\mathbf{w})$ may be regarded as a symmetric linear map from $\mathscr{E}_{\mathbf{w}} \wedge \mathscr{E}_{\mathbf{w}}$ to $\mathscr{E}_{\mathbf{w}}^* \wedge \mathscr{E}_{\mathbf{w}}^*$, so there are only 21 independent components. Then the condition (44.7) reduces the number by 1 to 20 independent components.

We define the *Ricci tensor* $\mathbf{S}$ by the component form

$$S_{\delta\beta} = R^{\alpha}{}_{\delta\beta\alpha}. \tag{44.10}$$

From (44.8) $\mathbf{S}$ is symmetric, so it has ten independent components. The trace of $\mathbf{S}$ with respect to σ is called the *scalar curvature*,

$$S = \operatorname{tr} \mathbf{S} = \Sigma^{\alpha\beta} S_{\alpha\beta}. \tag{44.11}$$

The combinations

$$G_{\alpha\beta} = S_{\alpha\beta} - \tfrac{1}{2} S \Sigma_{\alpha\beta} \tag{44.12}$$

define the components of the *Einstein tensor* $\mathbf{G}$, which is also symmetric and has ten independent components.

The curvature tensor $\mathbf{R}$ satisfies the *Bianchi identities*:

$$R^{\alpha}{}_{\delta\beta\gamma,\theta} + R^{\alpha}{}_{\delta\gamma\theta,\beta} + R^{\alpha}{}_{\delta\theta\beta,\gamma} = 0, \tag{44.13}$$

which may be verified by using (44.6) relative to a geodesic system at any point $\mathbf{w} \in \mathscr{E}$. (For the concept of a geodesic system see Section 57, IVT-2.) From (44.13) we may contract the pairs (α, γ) and (δ, β):

$$0 = S_{,\theta} + R^{\alpha\beta}{}_{\alpha\theta,\beta} + R^{\alpha\beta}{}_{\theta\beta,\alpha} = S_{,\theta} - 2S^{\beta}{}_{\theta,\beta} = -2G^{\beta}{}_{\theta,\beta}. \tag{44.14}$$

As we shall see, these contracted Bianchi identities are important in Einstein's formulation of the field equations for gravitation.

An important concept in Riemannian geometry is a geodesic (cf. Section 57, IVT-2). For a Minkowskian manifold we define a geodesic in a similar way: a curve $\boldsymbol{\lambda}(s)$ such that the tangent $\dot{\boldsymbol{\lambda}}$ forms a parallel vector field on $\boldsymbol{\lambda}$. The equations governing geodesics are

$$\frac{d^2\lambda^{\alpha}}{ds^2} + \begin{Bmatrix} \alpha \\ \beta\gamma \end{Bmatrix} \frac{d\lambda^{\beta}}{ds} \frac{d\lambda^{\gamma}}{ds} = 0. \tag{44.15}$$

We call a curve $\mathbf{w}(s)$ in $\mathscr{E}$ a *world line* if its tangent vector $\dot{\mathbf{w}}(s)$ is a timelike vector in $\mathscr{E}_{\mathbf{w}(s)}$ for all s. We may adjust the parameter s in such a way that

$$\sigma(\dot{\mathbf{w}}, \dot{\mathbf{w}}) = -c^2. \tag{44.16}$$

Then s is called the proper time as before.

Note. The tangent vector of a geodesic must have constant length with respect to the metric. Hence we can impose the condition (44.16) on a geodesic without affecting the geodesic equations (44.15).

As explained in Section 57, IVT-2, the geodesic equations correspond to the Euler–Lagrange equations for the arc length integral. This result may be applied to world lines in $\mathscr{E}$, and the integral is

$$I(\boldsymbol{\lambda}) = \int [-\sigma(\dot{\boldsymbol{\lambda}}, \dot{\boldsymbol{\lambda}})]^{1/2}\, ds. \tag{44.17}$$

Consequently, among all world lines joining two events the geodesic gives an extremum value of proper time interval between the events.

On any differentiable manifold we can define an exterior derivative operator d for differential forms. This operator is entirely independent of such metrical structure as the manifold may have. The definition and the basic properties of the operator d are explained in detail in Section 55, IVT-2. We recall that a differential form $\mathbf{A}$ is said to be *closed* if its exterior derivative $d\mathbf{A}$ vanishes:

$$d\mathbf{A} = \mathbf{0}. \tag{44.18}$$

From the lemma of Poincaré (cf. Section 52, IVT-2) the preceding condition is equivalent, locally, to the existence of a potential $\mathbf{B}$ such that

$$\mathbf{A} = d\mathbf{B}. \tag{44.19}$$

By Stokes' theorem (cf. Section 71, IVT-2) the condition (44.18) is also equivalent to the requirement that the integral of $\mathbf{A}$ over the boundary of any oriented domain $\mathscr{U}$ must vanish, viz.,

$$\int_{\partial\mathscr{U}} \mathbf{A} = \int_{\mathscr{U}} d\mathbf{A}, \tag{44.20}$$

where the dimension of $\partial\mathscr{U}$ is the same as the order of the differential form $\mathbf{A}$, of course.

Since we have assigned an orientation on the Minkowskian manifold $\mathscr{E}$, the metric σ determines a positive unit density tensor field $\boldsymbol{\Xi}$ which satisfies the condition

$$\sigma_4(\boldsymbol{\Xi}, \boldsymbol{\Xi}) = -1. \tag{44.21}$$

At each point $\mathbf{w} \in \mathscr{E}$ we define $\boldsymbol{\Xi}(\mathbf{w})$ by

$$\boldsymbol{\Xi}(\mathbf{w}) = \mathbf{e}^1(\mathbf{w}) \wedge \mathbf{e}^2(\mathbf{w}) \wedge \mathbf{e}^3(\mathbf{w}) \wedge \mathbf{e}^4(\mathbf{w}), \tag{44.22}$$

where $\{\mathbf{e}^\alpha(\mathbf{w})\}$ is any positive orthonormal basis in $\mathscr{E}_\mathbf{w}{}^*$. Relative to an arbitrary positive coordinate system (x^α), $\boldsymbol{\Xi}$ has the component form

$$\boldsymbol{\Xi} = (-\Sigma)^{1/2}\mathbf{h}^1 \wedge \mathbf{h}^2 \wedge \mathbf{h}^3 \wedge \mathbf{h}^4, \tag{44.23}$$

where Σ denotes the determinant of $[\Sigma_{\alpha\beta}]$:

$$\Sigma = \det[\Sigma_{\alpha\beta}], \tag{44.24}$$

which is negative for a Minkowskian metric.

Using the 4-form $\boldsymbol{\Xi}$, we define the integral of any function K (i.e., a 0-form) on an oriented domain $\mathscr{D}$ by

$$\int_{\mathscr{D}} K\boldsymbol{\Xi} = \int_{\mathscr{D}} K(-\Sigma)^{1/2}\, d\omega, \tag{44.25}$$

where $d\omega$ denotes the element of volume in (x^α); i.e., $d\omega = dx^1\, dx^2\, dx^3\, dx^4$. It is understood that (x^α) is a positive coordinate system on $\mathscr{D}$. Notice that the 4-form $K\boldsymbol{\Xi}$ is just the dual of the 0-form K, viz.,

$$K\boldsymbol{\Xi} = \mathbf{D}_0(K), \tag{44.26}$$

where the duality operator $\mathbf{D}_0$ is defined as before by (40.27). In particular, for any Lorentz frame $\{\mathbf{e}_\alpha(\mathbf{w})\}$ at $\mathbf{w}$

$$\mathbf{D}_0(1) = \mathbf{e}^1(\mathbf{w}) \wedge \mathbf{e}^2(\mathbf{w}) \wedge \mathbf{e}^3(\mathbf{w}) \wedge \mathbf{e}^4(\mathbf{w}) = \boldsymbol{\Xi}(\mathbf{w}), \quad \mathbf{D}_4(\boldsymbol{\Xi}) = -1, \tag{44.27}$$

where $\{\mathbf{e}^\alpha(\mathbf{w})\}$ denotes the dual basis of $\{\mathbf{e}_\alpha(\mathbf{w})\}$.

45. The Stress-Energy-Momentum Tensor in a Material Medium

In Section 41 we have shown that, in accord with the Lorentz electron theory, there is a symmetric stress-energy-momentum tensor $\boldsymbol{\Omega}$ associated with an electromagnetic field such that the divergence of $\boldsymbol{\Omega}$ is related to the momentum-energy production vector $\boldsymbol{\xi}$ by the equation of balance

$$\operatorname{Div} \boldsymbol{\Omega} + \boldsymbol{\xi} = \mathbf{0}. \tag{45.1}$$

Relative to a Lorentz frame $\{\mathbf{e}_i, \mathbf{e}_4\}$ the spacelike components ξ^i of $\boldsymbol{\xi}$ correspond to the components of the production of linear momentum, and the timelike component ξ^4 corresponds to $1/c$ times the production of energy due to the electromagnetic field. Relative to the same frame

$\{\mathbf{e}_i, \mathbf{e}_4\}$ the components Ω^{ij} correspond to the components of the Maxwell stress tensor, the components $\Omega^{i4} = \Omega^{4i}$ correspond to $1/c$ times the components of the Poynting vector, and the component Ω^{44} corresponds to the electromagnetic energy density of the field. We now show that a similar stress-energy-momentum tensor may be defined for a material medium.

In this formulation we temporarily use the structure of a Minkowskian space–time relative to which the divergence in (45.1) is taken. We recall first that in continuum mechanics a material medium has a mass density $\varrho(\mathbf{x}, t)$ and a symmetric stress tensor $\mathbf{T}(\mathbf{x}, t)$ such that

$$\frac{\partial \varrho}{\partial t} + \operatorname{div}(\varrho \mathbf{v}) = 0, \tag{45.2}$$

and

$$\operatorname{div} \mathbf{T} + \varrho \mathbf{b} = \varrho \mathbf{a}, \tag{45.3}$$

where $\mathbf{v}$ and $\mathbf{a}$ denote the velocity field and the acceleration field, respectively, relative to an inertial frame, and where $\mathbf{b}$ denotes the body force field. We regard the field $\varrho\mathbf{b}$ as a supply of linear momentum (i.e., the opposite of the production of linear momentum). Using (45.2) and Euler's formula (14.18), we can express the right-hand side of (45.3) as

$$\varrho \mathbf{a} = \frac{\partial(\varrho \mathbf{v})}{\partial t} + \operatorname{div}(\varrho \mathbf{v} \otimes \mathbf{v}). \tag{45.4}$$

Then the equation of linear momentum (45.3) may be rewritten as

$$\frac{\partial(\varrho \mathbf{v})}{\partial t} + \operatorname{div}(\varrho \mathbf{v} \otimes \mathbf{v} - \mathbf{T}) = \varrho \mathbf{b}. \tag{45.5}$$

This equation is quite similar to the equation of balance of linear momentum (33.21) in the classical theory of electromagnetism.

The equations of balance (45.2) and (45.5) may be revised to meet the invariance requirement of special relativity in the following way: First, a motion of the material medium corresponds to a collection of nonintersecting world lines in Minkowskian space–time. As explained in Section 39, we parametrize the world lines by the proper time s. Then the tangent vector field $\dot{\mathbf{w}}$ satisfies the condition of normalization (44.16). At each point $\mathbf{w}$ in the medium we put

$$\hat{\mathbf{e}}_4(\mathbf{w}) \equiv \frac{1}{c} \dot{\mathbf{w}}, \tag{45.6}$$

and we extend this timelike unit vector into a Lorentz frame $\{\hat{\mathbf{e}}_\alpha(\mathbf{w})\}$.

Now we define the stress tensor $\mathbf{T}(\mathbf{w})$ at the point $\mathbf{w}$ by the component form

$$\mathbf{T}(\mathbf{w}) = \hat{T}^{ij}(\mathbf{w})\hat{\mathbf{e}}_i(\mathbf{w}) \otimes \hat{\mathbf{e}}_j(\mathbf{w}), \tag{45.7}$$

where $\hat{T}^{ij}(\mathbf{w})$ denote the components of the usual Cauchy stress tensor relative to the spatial frame $\{\hat{\mathbf{e}}_i(\mathbf{w})\}$ which is formed by the first three vectors of the Lorentz frame $\{\hat{\mathbf{e}}_i, \hat{\mathbf{e}}_4\}$ defined at the point $\mathbf{w}$. From the condition (45.6) we know that the velocity vanishes at $\mathbf{w}$ relative to the Lorentz frame $\{\hat{\mathbf{e}}_i, \hat{\mathbf{e}}_4\}$. Consequently, the spatial frame $\{\hat{\mathbf{e}}_i\}$ is unique to within a rigid rotation only. As a result, the tensor $\mathbf{T}(\mathbf{w})$ is well defined by (45.7) and satisfies the condition

$$[\mathbf{T}(\mathbf{w})](\dot{\mathbf{w}}) = \mathbf{0}. \tag{45.8}$$

Relative to a Lorentz frame $\{\mathbf{e}_\alpha\}$ for the whole Minkowskian space–time the component form of the stress tensor field $\mathbf{T}$ is

$$\mathbf{T}(\mathbf{w}) = T^{\alpha\beta}(\mathbf{w})\mathbf{e}_\alpha \otimes \mathbf{e}_\beta, \tag{45.9}$$

where the components $T^{\alpha\beta}(\mathbf{w})$ may be obtained from the components $\hat{T}^{ij}(\mathbf{w})$ by a Lorentz transformation from $\{\mathbf{e}_\alpha\}$ to $\{\hat{\mathbf{e}}_i, \hat{\mathbf{e}}_4\}$. For example, suppose that the velocity at the particular point $\mathbf{w}$ happens to be $v\mathbf{e}_1$ relative to $\{\mathbf{e}_\alpha\}$. Then the transformation from $\{\mathbf{e}_\alpha\}$ to $\{\hat{\mathbf{e}}_\alpha(\mathbf{w})\}$ is given by

$$\begin{gathered}\hat{\mathbf{e}}_1(\mathbf{w}) = \frac{\mathbf{e}_1}{[1-(v/c)^2]^{1/2}} + \frac{(v/c)\mathbf{e}_4}{[1-(v/c)^2]^{1/2}}, \qquad \hat{\mathbf{e}}_2(\mathbf{w}) = \mathbf{e}_2, \\ \hat{\mathbf{e}}_3(\mathbf{w}) = \mathbf{e}_3, \qquad \hat{\mathbf{e}}_4(\mathbf{w}) = \frac{(v/c)\mathbf{e}_1}{[1-(v/c)^2]^{1/2}} + \frac{\mathbf{e}_4}{[1-(v/c)^2]^{1/2}}.\end{gathered} \tag{45.10}$$

Substituting (45.10) into (45.7) and then matching components with (45.9), we obtain

$$\begin{aligned} T^{11}(\mathbf{w}) &= \frac{\hat{T}^{11}(\mathbf{w})}{[1-(v/c)^2]^{1/2}}, & T^{12}(\mathbf{w}) &= \frac{\hat{T}^{12}(\mathbf{w})}{[1-(v/c)^2]^{1/2}}, \\ T^{13}(\mathbf{w}) &= \frac{\hat{T}^{13}(\mathbf{w})}{[1-(v/c)^2]^{1/2}}, & T^{14}(\mathbf{w}) &= \frac{(v/c)\hat{T}^{11}(\mathbf{w})}{[1-(v/c)^2]^{1/2}}, \\ T^{24}(\mathbf{w}) &= \frac{(v/c)\hat{T}^{12}(\mathbf{w})}{[1-(v/c)^2]^{1/2}}, & T^{34}(\mathbf{w}) &= \frac{(v/c)\hat{T}^{13}(\mathbf{w})}{[1-(v/c)^2]^{1/2}}, \\ T^{44}(\mathbf{w}) &= \frac{(v/c)^2\hat{T}^{11}(\mathbf{w})}{[1-(v/c)^2]^{1/2}}, & T^{22}(\mathbf{w}) &= \hat{T}^{22}(\mathbf{w}), \\ T^{23}(\mathbf{w}) &= \hat{T}^{23}(\mathbf{w}), & T^{33}(\mathbf{w}) &= \hat{T}^{33}(\mathbf{w}). \end{aligned} \tag{45.11}$$

Following the same procedure, we get the component form for $\dot{\mathbf{w}}$:

$$\dot{\mathbf{w}} = c\hat{\mathbf{e}}_4(\mathbf{w}) = \frac{v\mathbf{e}_1}{[1-(v/c)^2]^{1/2}} + \frac{c\mathbf{e}_4}{[1-(v/c)^2]^{1/2}}. \tag{45.12}$$

The fact that (45.11) and (45.12) are consistent with the condition (45.8) may be verified easily by direct calculation.

Next, the momentum-energy supply vector $\boldsymbol{\eta}(\mathbf{w})$ is defined in a similar way. Specifically, we use the special Lorentz frame $\{\hat{\mathbf{e}}_\alpha(\mathbf{w})\}$ at each point $\mathbf{w}$, and we define $\boldsymbol{\eta}(\mathbf{w})$ by the component form:

$$\boldsymbol{\eta}(\mathbf{w}) = \hat{\varrho}(\mathbf{w})\hat{b}^i(\mathbf{w})\hat{\mathbf{e}}_i(\mathbf{w}), \tag{45.13}$$

where $\hat{b}^i(\mathbf{w})$ denote the components of the body force relative to the spatial frame $\{\hat{\mathbf{e}}_i(\mathbf{w})\}$, and where $\hat{\varrho}(\mathbf{w})$ denotes a positive scalar, called the *mass density* of the material at the point $\mathbf{w}$.

Note. In the theory of relativity mass and energy are interchangeable quantities. Hence $\hat{\varrho}(\mathbf{w})$ includes both the proper mass and the internal energy of the medium. In the theory of some special classes of materials, e.g., the theory of hyperelastic materials, $\hat{\varrho}$ may be expressed as $\varrho(1 + \varepsilon/c^2)$, where ϱ is called the *proper mass density*, and where ε is called the *internal energy density* (per unit proper mass). In these models the proper mass density is treated as an intrinsic property of the material medium like the mass density in the classical theory of continuum mechanics. Then it is assumed that the proper mass satisfies a conservation law which forbids it from changing into energy. The factor $1/c^2$ for the term ε/c^2 has the usual meaning: the mass equivalent to the energy ε is ε/c^2. By virtue of the factor $1/c^2$ the value of ε/c^2 is generally much smaller than the value of ϱ. Hence $\hat{\varrho}$ and ϱ are almost the same.

Now by virtue of the remark about the uniqueness of the spatial frame $\{\hat{\mathbf{e}}_i(\mathbf{w})\}$ the vector $\boldsymbol{\eta}(\mathbf{w})$ is well defined by (45.13) and satisfies the condition

$$\sigma(\boldsymbol{\eta}, \dot{\mathbf{w}}) = 0. \tag{45.14}$$

If we use a Lorentz frame $\{\mathbf{e}_\alpha\}$ for the whole Minkowskian space–time $\mathscr{E}$, then the vector $\boldsymbol{\eta}(\mathbf{w})$ has the component form

$$\boldsymbol{\eta}(\mathbf{w}) = \eta^\alpha(\mathbf{w})\mathbf{e}_\alpha, \tag{45.15}$$

where the components $\eta^\alpha(\mathbf{w})$ may be obtained from the components $\hat{\eta}^i(\mathbf{w})$

by a Lorentz transformation from $\{\mathbf{e}_\alpha\}$ to $\{\hat{\mathbf{e}}_\alpha(\mathbf{w})\}$. For the special transformation (45.10)

$$\eta^1(\mathbf{w}) = \frac{\hat{\varrho}(\mathbf{w})\hat{b}^1(\mathbf{w})}{[1-(v/c)^2]^{1/2}}, \qquad \eta^2(\mathbf{w}) = \hat{\varrho}(\mathbf{w})\hat{b}^2(\mathbf{w}),$$
$$\eta^3(\mathbf{w}) = \hat{\varrho}(\mathbf{w})\hat{b}^3(\mathbf{w}), \qquad \eta^4(\mathbf{w}) = \frac{(v/c)\hat{\varrho}(\mathbf{w})\hat{b}^1(\mathbf{w})}{[1-(v/c)^2]^{1/2}}. \tag{45.16}$$

As before we can verify that (45.16) and (45.12) are consistent with (45.14).

Now we define the stress-energy-momentum tensor $\mathbf{\Theta}$ by

$$\mathbf{\Theta} = \hat{\varrho}\dot{\mathbf{w}} \otimes \dot{\mathbf{w}} - \mathbf{T}. \tag{45.17}$$

Clearly $\mathbf{\Theta}$ is a symmetric tensor field on the domain of the medium. We claim that this tensor field satisfies the equation of balance

$$\operatorname{Div} \mathbf{\Theta} - \boldsymbol{\eta} = \mathbf{0}. \tag{45.18}$$

Notice that the minus sign of $\boldsymbol{\eta}$ is due to the fact that $\boldsymbol{\eta}$ denotes the supply of momentum-energy, while $\boldsymbol{\xi}$ in (45.1) denotes the production of momentum-energy in the medium.

To prove the equation of balance (45.18), we choose as the Lorentz frame $\{\mathbf{e}_\alpha\}$ the special frame $\{\hat{\mathbf{e}}_\alpha(\mathbf{w})\}$ at a particular point $\mathbf{w}$, and we verify the validity of (45.18) at the point $\mathbf{w}$. The components of (45.18) in a Lorentz system (x^i, ct) in general are

$$\frac{\partial}{\partial x^j}(\hat{\varrho}\dot{x}^i\dot{x}^j - T^{ij}) + \frac{1}{c}\frac{\partial}{\partial t}(\hat{\varrho}\dot{x}^i c\dot{t} - T^{i4}) = \eta^i,$$
$$\frac{\partial}{\partial x^j}(\hat{\varrho}c\dot{t}\dot{x}^j - T^{4j}) + \frac{1}{c}\frac{\partial}{\partial t}(\hat{\varrho}c^2\dot{t}^2 - T^{44}) = \eta^4. \tag{45.19}$$

From (45.6) in the special Lorentz frame $\{\hat{\mathbf{e}}_\alpha(\mathbf{w})\}$ at the point $\mathbf{w}$ the components $\dot{x}^i(\mathbf{w})$ vanish and the component $c\dot{t}$ reduces to c. Also, the components $T^{i4}(\mathbf{w})$, $T^{44}(\mathbf{w})$, and $\eta^4(\mathbf{w})$ all vanish; cf. (45.7) and (45.13).

From the general conditions (39.19) and (39.21) we have

$$\left.\frac{\partial \dot{x}^i}{\partial x^j}\right|_{\mathbf{w}} = \left.\frac{\partial v^i}{\partial x^j}\right|_{\mathbf{w}}, \quad \left.\frac{\partial \dot{x}^i}{\partial t}\right|_{\mathbf{w}} = \left.\frac{\partial v^i}{\partial t}\right|_{\mathbf{w}}, \quad \left.\frac{\partial \dot{t}}{\partial x^j}\right|_{\mathbf{w}} = \left.\frac{\partial \dot{t}}{\partial t}\right|_{\mathbf{w}} = 0, \tag{45.20}$$

at the particular point $\mathbf{w}$. Then from (45.8) we obtain

$$T^{ij}\dot{x}^j - T^{i4}c\dot{t} = 0, \qquad T^{4j}\dot{x}^j - T^{44}c\dot{t} = 0. \tag{45.21}$$

Differentiating these conditions with respect to x^i and t and evaluating the results at the point $\mathbf{w}$, we get

$$T^{ij}(\mathbf{w}) \frac{\partial v^j}{\partial x^i}\bigg|_{\mathbf{w}} = c \frac{\partial T^{i4}}{\partial x^i}\bigg|_{\mathbf{w}}, \qquad T^{ij}(\mathbf{w}) \frac{\partial v^j}{\partial t}\bigg|_{\mathbf{w}} = c \frac{\partial T^{i4}}{\partial t}\bigg|_{\mathbf{w}},$$
$$\frac{\partial T^{44}}{\partial x^i}\bigg|_{\mathbf{w}} = 0, \qquad \frac{\partial T^{44}}{\partial t}\bigg|_{\mathbf{w}} = 0. \tag{45.22}$$

Substituting (45.20) and (45.22) into (45.19), we see that

$$-\frac{\partial T^{ij}}{\partial x^j}\bigg|_{\mathbf{w}} + \frac{\partial(\hat{\varrho} v^i)}{\partial t}\bigg|_{\mathbf{w}} - \frac{1}{c^2} T^{ij}(\mathbf{w}) \frac{\partial v^j}{\partial t}\bigg|_{\mathbf{w}} = \hat{\varrho} b^i, \tag{45.23a}$$

$$c \frac{\partial(\hat{\varrho} v^j)}{\partial x^j}\bigg|_{\mathbf{w}} - \frac{1}{c} T^{ij}(\mathbf{w}) \frac{\partial v^j}{\partial x^i}\bigg|_{\mathbf{w}} + c \frac{\partial \hat{\varrho}}{\partial t}\bigg|_{\mathbf{w}} = 0. \tag{45.23b}$$

Now (45.23a) is just the classical equation of linear momentum (45.5) evaluated at the particular point $\mathbf{w}$, where the velocity $\mathbf{v}$ vanishes, except that the mass density $\hat{\varrho}$ here contains the internal energy, and that an extra relativistic correction term $-(1/c^2)T^{ij}(\mathbf{w})(\partial v^j/\partial t)|_{\mathbf{w}}$ is included. That term is due to the fact that $T^{ij}v^j\mathbf{e}_i$ may be regarded as the energy flux induced by the stress and, thus, $(1/c^2)T^{ij}v^j\mathbf{e}_i$ is its corresponding momentum flux. Therefore the time derivative $(1/c^2)T^{ij}(\mathbf{w})(\partial v^j/\partial t)|_{\mathbf{w}}$ is a momentum production vector.

Next, dividing (45.23b) by c, we get

$$\frac{\partial(\hat{\varrho} v^j)}{\partial x^j}\bigg|_{\mathbf{w}} - \frac{1}{c^2} T^{ij}(\mathbf{w}) \frac{\partial v^j}{\partial x^i}\bigg|_{\mathbf{w}} + \frac{\partial \hat{\varrho}}{\partial t}\bigg|_{\mathbf{w}} = 0. \tag{45.24}$$

This is just the classical continuity equation (45.2) evaluated at the point $\mathbf{w}$, except that ϱ is replaced by $\hat{\varrho}$, and that an extra relativistic correction term $-(1/c^2)T^{ij}(\mathbf{w})(\partial v^j/\partial x^i)|_{\mathbf{w}}$ is included. That term is due to the fact that $T^{ij}(\partial v^i/\partial x^j)$ is the stress power. Hence $1/c^2$ times it, $(1/c^2)T^{ij}(\mathbf{w}) \times (\partial v^i/\partial x^j)|_{\mathbf{w}}$, is the corresponding rate of mass production at the point $\mathbf{w}$. If we regard the mass density $\hat{\varrho}$ as $\varrho(1 + \varepsilon/c^2)$, then the proper mass density ϱ satisfies the continuity equation (45.2), and (45.24) reduces to the classical energy equation (20.8).

The preceding analysis shows that relative to the special Lorentz frame $\{\mathbf{e}_\alpha\} = \{\hat{\mathbf{e}}_\alpha(\mathbf{w})\}$, (45.18) is the correct equation of balance at the point $\mathbf{w}$. However, since (45.18) is a tensor equation, it is independent of any choice of Lorentz frame. Thus the proof is complete.

It follows from the equation of balance (45.18) that relative to any Lorentz frame $\{\mathbf{e}_\alpha\}$ the spacelike component $(\partial\Theta^{i\alpha}/\partial x^\alpha)\mathbf{e}_i$ corresponds to the supply of momentum, while the timelike component $(\partial\Theta^{4\alpha}/\partial x^\alpha)\mathbf{e}_4$ corresponds to $1/c$ times the supply of energy as observed in $\{\mathbf{e}_\alpha\}$.

Note. The stress-energy-momentum tensor $\mathbf{\Theta}$ defined by (45.17) satisfies the condition

$$[\mathbf{\Theta}(\mathbf{w})](\dot{\mathbf{w}}) = -\frac{\hat{\varrho}}{c^2}\,\dot{\mathbf{w}}; \tag{45.25}$$

i.e., $\dot{\mathbf{w}}$ is a proper vector of $\mathbf{\Theta}(\mathbf{w})$ corresponding to the proper number $-\hat{\varrho}/c^2$. The preceding result is due to the fact that in the definition (45.7) for the tensor $\mathbf{T}(\mathbf{w})$ we have not allowed any extra momentum-energy flux such as a heat flux and a heat supply.

If the stress tensor $\mathbf{T}$ always vanishes in the medium, and if there is no internal energy ε, then the stress-energy-momentum tensor is just the simple tensor $\varrho\dot{\mathbf{w}} \otimes \dot{\mathbf{w}}$. Such a medium is called an *incoherent fluid* or a *dust cloud.* In general we call $\varrho\dot{\mathbf{w}} \otimes \dot{\mathbf{w}}$ the *intrinsic stress-energy-momentum* and $(\varrho\varepsilon/c^2)\dot{\mathbf{w}} \otimes \dot{\mathbf{w}} - \mathbf{T}$ the *extra stress-energy-momentum* of the material medium. Their sum is the stress-energy-momentum $\mathbf{\Theta}$.

Note. Since the tangent vector $\dot{\mathbf{w}}$ must satisfy the condition of normalization (44.16), the trace of the intrinsic stress-energy-momentum is $\varrho\sigma(\dot{\mathbf{w}}, \dot{\mathbf{w}}) = -\varrho c^2$, which is just the negative of the energy density of the proper mass of the medium. This energy density is generally much greater than the trace $-\varrho\varepsilon - \hat{T}^{11} - \hat{T}^{22} - \hat{T}^{33}$ of the extra stress-energy-momentum. In this sense we say that the intrinsic stress-energy-momentum is the *dominant part* of $\mathbf{\Theta}$.

So far we have defined a stress-energy-momentum tensor $\mathbf{\Theta}(\mathbf{w})$ at each point $\mathbf{w}$ in a material medium by using the structure of the Minkowskian space–time in special relativity. We have shown that the tensor field $\mathbf{\Theta}$ satisfies the equation of balance (45.18), which has the same form as the equation of balance (45.1). If we regard a Lorentz system in the special theory of relativity as the counterpart of an inertial system in the classical theory, then the equations of balance (45.18) and (45.1) are just the relativistic versions of the classical equations of balance of mass and momentum.

The formulation of the stress-energy-momentum in a material medium may be generalized in an obvious way from special relativity to general relativity. We simply regard the frame $\{\mathbf{e}_\alpha(\mathbf{w})\}$ as being a Lorentz basis at the particular point $\mathbf{w}$ in the Minkowskian manifold. The condition (45.6) still determines the timelike basis vector $\hat{\mathbf{e}}_4(\mathbf{w})$ uniquely, and then the spacelike basis vectors $\{\hat{\mathbf{e}}_i(\mathbf{w})\}$ are unique to within an arbitrary rotation as before. Of course, we still define $\mathbf{T}(\mathbf{w})$ by (45.7) and $\mathbf{\Theta}(\mathbf{w})$ by (45.17).

The definition (45.17) is suitable for a medium which has no electromagnetic fields. If electromagnetic fields are present, then we modify the definition of $\mathbf{\Theta}$ to

$$\mathbf{\Theta} = \hat{\varrho}\dot{\mathbf{w}} \otimes \dot{\mathbf{w}} - \mathbf{T} + \mathbf{\Omega}, \tag{45.26}$$

where $\mathbf{\Omega}$ denotes the stress-energy-momentum tensor defined by the electromagnetic fields. [We shall formulate the theory of electromagnetism in the context of general relativity in Chapter 8. The definition of the stress-energy-momentum tensor $\mathbf{\Omega}$ is given by (55.1) which is a direct generalization of a similar definition (41.19) in special relativity.]

Note. The definition of the stress-energy-momentum tensor $\mathbf{\Theta}$ must be modified further when other kinds of energy and momentum fluxes are present, such as heat conduction. We shall not consider such other fluxes in this treatise, however.

In the view of Einstein there is a distribution of stress-energy-momentum tensor on the Minkowskian manifold $\mathscr{E}$ in the general theory of relativity. According to Einstein's formulation, the field of stress-energy-momentum tensors coincides with the field of Einstein tensors associated with the Minkowskian metric σ. Consequently, in the general theory the model for the event world is not completely determined *a priori* as in the special theory and in the classical theory. Indeed, the structure of the particular Minkowskian manifold, which represents the event world, depends on such entities as mass, stress, electromagnetic field, etc. which may be introduced into the manifold. This feature of the general theory renders it impossible to treat rigorously the entities separately, since an overall interaction of the entities is introduced into the model at the onset of the formulation. Fortunately, the stress-energy-momentum tensor may be decomposed into a dominant part and an extra part. To the extent that the extra part does not alter the structure of the Minkowskian manifold significantly, we develop separately in the context of general relativity a theory of mechanics in this chapter and a theory of electromagnetism in the following chapter. An example of electromagneto–gravitational interaction is treated in the last section.

46. Einstein's Field Equations

In accord with Einstein's general theory of relativity the stress-energy-momentum distribution on $\mathscr{E}$ gives rise to the departure of the Minkowskian manifold from the Minkowskian (affine) space–time. He wrote down the field equations:

$$G^{\alpha\beta} - C\Sigma^{\alpha\beta} = -K\Theta^{\alpha\beta}, \tag{46.1}$$

which are now known as *Einstein's field equations*. In these equations C and K are two physical constants, which we shall identify later. Since Σ is covariantly constant, and since **G** satisfies the contracted Bianchi identities (44.14), it follows directly from (46.1) that we must have

$$[\mathrm{Div}\,\Theta]^{\alpha} = \Theta^{\alpha\beta}{}_{,\beta} = 0, \tag{46.2}$$

where the divergence is taken with respect to the Minkowskian metric σ.

Comparing (46.2) with the special relativistic equation (45.18), we see that the momentum-energy supply vector $\boldsymbol{\eta}$ is absorbed entirely by the curvature of the Minkowskian metric. Since the components of the Einstein tensor **G** may be expressed in terms of the second partial derivatives of the components of the Minkowskian metric σ, the field equations (46.1) are just the relativistic version of the Poisson equation (43.9) with the metric corresponding to the gravitational potential ζ and the stress-energy-momentum corresponding to the mass density ϱ.

In application the precise value of the stress-energy-momentum is not easy to determine. The prevailing view in physics is that proper mass, like electric charge, is confined to some small bounded regions. Outside these regions the system of field equations (46.1) reduces to the system of homogeneous equations

$$G^{\alpha\beta} - C\Sigma^{\alpha\beta} = 0. \tag{46.3}$$

In the small bounded regions containing some proper mass, $\mathbf{\Sigma}$ and **G** may become singular.

We may regard the stress-energy-momentum tensor $\mathbf{\Theta}$ simply as an average of $(\mathbf{G} - C\mathbf{\Sigma})/K$ over the singularities. This interpretation is similar to that of the concept of a density in the classical Newtonian theory. Specifically, we regard the Newtonian gravitational potential ζ as a solution of the homogeneous field equation

$$\nabla^2\zeta = 0 \tag{46.4}$$

except on some small domains occupied by mass where ζ may become singular. Then the mass density ϱ is just the average of $\nabla^2\zeta/4\pi k$ over the singularities.

We notice that the field equations (46.1) and (46.3) contain a term $C\Sigma^{\alpha\beta}$, which has no counterpart in the classical theory. If we require that $\mathscr{E}$ approach a flat Minkowskian (affine) space–time "at infinity," then we must set $C = 0$. The prevailing view among cosmologists, however, is that the spatial slices of $\mathscr{E}$ are, in fact, finite. Hence it is meaningless to talk about the structure of $\mathscr{E}$ "at infinity." In other words a spatial slice of $\mathscr{E}$ does not resemble an infinite affine space at large with islands of material media floating on it causing curvature in adjacent regions. Rather, each spatial slice has a finite volume only, and there is an average nonzero density of proper mass on such a slice.

One possible solution of (46.3) is given by a Minkowskian manifold with constant curvature, called a *de Sitter manifold*, such that

$$\begin{gathered} R_{\alpha\beta\gamma\delta} = \frac{C}{3}\,(\Sigma_{\alpha\gamma}\Sigma_{\beta\delta} - \Sigma_{\alpha\delta}\Sigma_{\beta\gamma}), \\ S_{\beta\gamma} = -C\Sigma_{\beta\gamma}, \qquad G_{\beta\gamma} = C\Sigma_{\beta\gamma}, \qquad S = -4C. \end{gathered} \tag{46.5}$$

In that manifold the Gaussian curvatures on geodesic surfaces are independent of the choice of the surface at each point of the manifold. When $C > 0$ in (46.5), all spacelike geodesics starting from any point will meet again at a certain point, possibly the starting point itself, while all timelike geodesics will open out and never meet again. When $C < 0$, the behavior of the geodesics is just the opposite of that in the preceding case.

Einstein introduced the constant C in order to discuss extremely large-scale problems in cosmology. It is generally agreed that the value of that cosmologic constant is extremely small. Hence we shall neglect its contribution to the field equations. Thus we write

$$G^{\alpha\beta} = -K\Theta^{\alpha\beta}, \tag{46.6}$$

and we proceed to determine the value of the constant K in terms of the gravitational constant k.

First, according to Einstein's interpretation of the Minkowskian manifold $\mathscr{E}$, the world line of a particle is a geodesic with respect to the Minkowskian metric σ. In the special theory of relativity $\mathscr{E}$ is a Minkowskian space–time, which is an affine space equipped with a constant Minkowskian metric. Then there is no curvature, and a geodesic is just a straight line. That metric corresponds to the zeroth-order approximate solution to the field equations (46.6).

Next, in the first-order approximation, we take $[\Sigma_{\alpha\beta}]$ to be a small departure from the constant matrix $[\Delta_{\alpha\beta}] = \operatorname{diag}(1, 1, 1, -1)$ by a variable field $[F_{\alpha\beta}]$, such that all terms higher than the first order in $\mathbf{F}$ may be neglected. Also, the motion is assumed to have a slow speed compared to the speed of light, so that $\dot{x}^i = v^i\dot{t}$ is neglible relative to $\dot{x}^4 = c\dot{t}$. Under these assumptions the system of geodesic equations (44.15) reduces to

$$\frac{d^2x^i}{ds^2} = \frac{\partial}{\partial x^i}\left(\frac{1}{2}\,c^2F_{44}\right), \qquad i = 1, 2, 3. \tag{46.7}$$

The system (46.7) is similar to Newton's law of gravitation with ζ given by

$$\zeta = -\frac{1}{2}\,c^2F_{44}. \tag{46.8}$$

Note. The factor C^2 in (46.7) and (46.8) is due to the fact that the Minkowskian metric σ has a component matrix approximately equal to the constant matrix $[\Delta_{\alpha\beta}] = \operatorname{diag}(1, 1, 1, -1)$ in the coordinate system $(x^\alpha) = (x^i, ct)$. Such a coordinate system corresponds approximately to a Lorentz system in the special theory and an inertial system in the classical theory. Some authors prefer to use the metric $\hat{\sigma} \equiv -(1/c^2)\sigma$ on $\mathscr{E}$. Then the coordinate system $(\hat{x}^\alpha)$ such that the component matrix $[\hat{\Sigma}_{\alpha\beta}]$ is approximately equal to the constant matrix $[-\Delta_{\alpha\beta}] = \operatorname{diag}(-1, -1, -1, 1)$ is related to our coordinate system (x^α) by $(\hat{x}^\alpha) = (x^\alpha/c) = (x^i/c, t)$, where (x^i, t) corresponds approximately to an inertial system in the classical theory. In the coordinate system $(\hat{x}^\alpha)$ we have $\dot{\hat{x}}^i = (v^i/c)\dot{t}$ and $\dot{\hat{x}}^4 = \dot{t}$. Then the condition of normalization (44.16) becomes $\hat{\sigma}(\dot{\mathbf{w}}, \dot{\mathbf{w}}) = 1$, and the system of geodesic equations (46.7) becomes

$$\frac{d^2\hat{x}^i}{ds^2} = \frac{\partial}{\partial \hat{x}^i}\left(\frac{1}{2}\,\hat{F}_{44}\right). \tag{46.9}$$

Although some equations in the general theory take slightly simpler forms in terms of the metric $\hat{\sigma}$ and the coordinate system $(\hat{x}^\alpha)$, we prefer to use the metric σ and the coordinate system $(x^\alpha) = (x^i, ct)$, since they are directly comparable to corresponding quantities in the special theory and the classical theory.

Now taking the trace of the field equations (46.6) and using the definition (44.12) for the Einstein tensor $\mathbf{G}$, we get

$$S = K\Theta, \tag{46.10}$$

where Θ denotes the trace of $\mathbf{\Theta}$, viz.,

$$\Theta = \Sigma_{\alpha\beta}\Theta^{\alpha\beta}. \tag{46.11}$$

Substituting (46.10) into (46.6), we have

$$S_{\alpha\beta} = -K(\Theta_{\alpha\beta} - \tfrac{1}{2}\Sigma_{\alpha\beta}\Theta), \tag{46.12a}$$

$$\equiv -K\Theta^*_{\alpha\beta}. \tag{46.12b}$$

In the first-order approximation the Ricci tensor $\mathbf{S}$ is given by

$$\begin{aligned} S_{\delta\beta} &= \frac{\partial}{\partial x^\beta}\begin{Bmatrix}\alpha\\ \delta\alpha\end{Bmatrix} - \frac{\partial}{\partial x^\alpha}\begin{Bmatrix}\alpha\\ \delta\beta\end{Bmatrix} + \begin{Bmatrix}\lambda\\ \delta\alpha\end{Bmatrix}\begin{Bmatrix}\alpha\\ \lambda\beta\end{Bmatrix} - \begin{Bmatrix}\lambda\\ \delta\beta\end{Bmatrix}\begin{Bmatrix}\alpha\\ \lambda\alpha\end{Bmatrix} \\ &= \frac{1}{2}\Delta^{\alpha\mu}\left(\frac{\partial^2 F_{\alpha\mu}}{\partial x^\delta\,\partial x^\beta} + \frac{\partial^2 F_{\beta\delta}}{\partial x^\alpha\,\partial x^\mu} - \frac{\partial^2 F_{\mu\beta}}{\partial x^\alpha\,\partial x^\delta} - \frac{\partial^2 F_{\mu\delta}}{\partial x^\beta\,\partial x^\alpha}\right) \\ &= \frac{1}{2}\Delta^{\alpha\mu}\frac{\partial^2 F_{\beta\delta}}{\partial x^\alpha\,\partial x^\mu} - \frac{1}{2}\frac{\partial}{\partial x^\delta}\left(\frac{\partial F_\beta{}^{*\alpha}}{\partial x^\alpha}\right) - \frac{1}{2}\frac{\partial}{\partial x^\beta}\left(\frac{\partial F_\delta{}^{*\alpha}}{\partial x^\alpha}\right), \end{aligned} \tag{46.13}$$

where

$$F_\beta{}^{*\alpha} \equiv \Delta^{\alpha\mu}F_{\mu\beta} - \tfrac{1}{2}\Delta^{\sigma\tau}F_{\sigma\tau}\delta_\beta{}^\alpha. \tag{46.14}$$

By using a coordinate transformation which preserves the first-order approximation, we can impose four more conditions on the components $F_{\alpha\beta}$ of the metric. Specifically, without loss of generality, we may assume that

$$\frac{\partial F_\beta{}^{*\alpha}}{\partial x^\alpha} = 0, \qquad \beta = 1, 2, 3, 4. \tag{46.15}$$

Under these conditions the formula (46.13) for $S_{\delta\beta}$ is simplified a great deal.

In order to satisfy the coordinate conditions (46.15), we change the coordinate system (x^α) to the system $(\bar{x}^\alpha)$ by a transformation of the form

$$\bar{x}^\alpha = x^\alpha + z^\alpha(x^\beta), \tag{46.16}$$

where $z^\alpha(x^\beta)$ are certain undetermined functions of the first order with respect to $\mathbf{F}$. To within first-order accuracy the inverse transformation of (46.16) is given by

$$x^\alpha = \bar{x}^\alpha - z^\alpha(\bar{x}^\beta). \tag{46.17}$$

From the usual transformation law of the components of a tensor we then have

$$\Delta_{\mu\nu} + \bar{F}_{\mu\nu} = \left(\delta_\mu{}^\alpha - \frac{\partial z^\alpha}{\partial x^\mu}\right)\left(\delta_\nu{}^\beta - \frac{\partial z^\beta}{\partial x^\nu}\right)(\Delta_{\alpha\beta} + F_{\alpha\beta}), \tag{46.18}$$

which implies

$$\bar{F}_{\mu\nu} = F_{\mu\nu} - \varDelta_{\alpha\mu}\frac{\partial z^\alpha}{\partial x^\nu} - \varDelta_{\alpha\nu}\frac{\partial z^\alpha}{\partial x^\mu}. \tag{46.19}$$

It follows that the transformation law of the components of the tensor $\mathbf{F}^*$ is

$$\bar{F}_\beta{}^{*\alpha} = F_\beta{}^{*\alpha} - \frac{\partial z^\alpha}{\partial x^\beta} - \varDelta^{\alpha\mu}\varDelta_{\gamma\beta}\frac{\partial z^\gamma}{\partial x^\mu} + \frac{\partial z^\mu}{\partial x^\mu}\delta_\beta{}^\alpha. \tag{46.20}$$

Differentiating this equation with respect to $\bar{x}^\alpha$, we obtain

$$\frac{\partial \bar{F}_\beta{}^{*\alpha}}{\partial \bar{x}^\alpha} = \frac{\partial F_\beta{}^{*\alpha}}{\partial x^\alpha} - \varDelta^{\alpha\mu}\varDelta_{\gamma\beta}\frac{\partial^2 z^\gamma}{\partial x^\mu\,\partial x^\alpha}. \tag{46.21}$$

Then

$$\frac{\partial \bar{F}_\beta{}^{*\alpha}}{\partial \bar{x}^\alpha} = 0, \qquad \beta = 1, 2, 3, 4, \tag{46.22}$$

provided that the undetermined functions z^α satisfy the partial differential equations

$$\varDelta^{\alpha\mu}\frac{\partial^2 z^\gamma}{\partial x^\alpha\,\partial x^\mu} = \varDelta^{\beta\gamma}\frac{\partial F_\beta{}^{*\alpha}}{\partial x^\alpha}, \qquad \gamma = 1, 2, 3, 4, \tag{46.23}$$

which may be solved by the retarded potentials; cf. (32.19) and (32.21). Thus a coordinate system $(\bar{x}^\alpha)$ in which the coordinate conditions (46.22) are satisfied exists.

In the special coordinate system (x^α) such that (46.15) holds, the formula (46.13) simplifies to

$$S_{\delta\beta} = \frac{1}{2}\varDelta^{\alpha\mu}\frac{\partial^2 F_{\beta\delta}}{\partial x^\alpha\,\partial x^\mu}. \tag{46.24}$$

Now in the first-order approximation the derivative with respect to x^4 is small compare to that with respect to a spatial variable x^i; i.e., the components of $\mathbf{F}$ may be regarded as approximately independent of x^4. Then the right-hand side of (46.24) reduces to the ordinary (three-dimensional) Laplacian of $F_{\beta\delta}$, viz.,

$$S_{\delta\beta} = \tfrac{1}{2}\nabla^2 F_{\beta\delta}. \tag{46.25}$$

In particular, for the component F_{44}

$$S_{44} = \tfrac{1}{2}\nabla^2 F_{44}. \tag{46.26}$$

Using the field equations (46.6), we then have

$$-K\Theta^*_{44} = \tfrac{1}{2}\nabla^2 F_{44}. \tag{46.27}$$

In the first-order approximation the stress-energy-momentum tensor Θ is dominated by the (4, 4) component of the intrinsic part $\varrho\dot{\mathbf{w}} \otimes \dot{\mathbf{w}}$, viz.,

$$\Theta^{44} = \Theta_{44} = \varrho c^2, \qquad \Theta = -\varrho c^2. \tag{46.28}$$

Consequently, Θ^*_{44} has the approximate value

$$\Theta^*_{44} = -\tfrac{1}{2}\varrho c^2. \tag{46.29}$$

Substituting this value into the Poisson equation (46.27) and comparing the result with (46.8) and (43.9), we obtain

$$K = \frac{8\pi k}{c^4} = 2.07 \times 10^{-48}\ \mathrm{cm}^{-1}\ \mathrm{g}^{-1}\ \mathrm{sec}^2, \tag{46.30}$$

where the numerical value of K in cgs units is obtained from that of k as shown in (43.2).

Note. In cgs units the unit of the constant K is $\mathrm{cm}^{-1}\ \mathrm{g}^{-1}\ \mathrm{sec}^2$, as shown in (46.30). It can be verified easily that this unit is consistent with the field equations (46.6). Indeed, in our formulation the unit of the components of the stress-energy-momentum tensor is $\mathrm{cm}^{-1}\ \mathrm{g}\ \mathrm{sec}^{-2}$ (i.e., $\mathrm{dyn/cm}^2$) and the unit of the components of the Einstein tensor is cm^{-2}; the latter unit follows from the fact that the unit of the coordinates $(x^\alpha) = (x^i, ct)$ is cm.

In the following section we shall summarize a rigorous solution of the homogeneous field equations

$$G^{\alpha\beta} = 0 \tag{46.31}$$

such that $\Sigma^{\alpha\beta}$, and therefore $G^{\alpha\beta}$, both become singular on a small bounded region. This solution was obtained originally by Schwarzschild. It is generally agreed that the Schwarzschild solution corresponds to the gravitational potential associated with a single particle according to the general theory of relativity.

47. The Schwarzschild Solution and the Problems of Planetary Orbits and the Deflection of Light

Einstein's field equations form a system of nonlinear second-order partial differential equations for the components of the Minkowskian metric when the components of the stress-energy-momentum tensor field are given. These equations are very difficult to solve, and only a few exact solutions are known. In this section we shall summarize one particular solution of the system of geometric equations

$$S_{\alpha\beta} = 0, \tag{47.1}$$

which is equivalent to the system of homogeneous field equations

$$G_{\alpha\beta} = 0 \tag{47.2}$$

by virtue of (44.12). The particular solution has two singularities near a given point $\mathbf{w} \in \mathscr{E}$ and may be regarded as the metric in the neighborhood of an isolated particle. The solution was discovered originally by Schwarzschild. One of the singularities is entirely different from the singularity in the classical theory.

The starting point for the derivation of the Schwarzschild solution is the assumption that the components of the metric be of the special form

$$\Sigma_{11} = e^{\lambda(r)}, \qquad \Sigma_{22} = r^2, \qquad \Sigma_{33} = r^2 \sin^2\theta, \qquad \Sigma_{44} = -e^{\nu(r)}, \tag{47.3}$$

where $\lambda(r)$ and $\nu(r)$ are certain undetermined functions of r. This form of metric corresponds to a gravitational potential which is spherically symmetric with respect to a particular world line, $r = 0$, in the coordinate system $(x^\alpha) = (r, \theta, \varphi, ct)$, where (r, θ, φ) are the spherical coordinates centered at the world line.

It can be shown that the nonzero Christoffel symbols based on the metric given by (47.3) are

$$\begin{gathered}
\begin{Bmatrix} 1 \\ 11 \end{Bmatrix} = \frac{1}{2}\lambda', \quad \begin{Bmatrix} 1 \\ 22 \end{Bmatrix} = -re^{-\lambda}, \quad \begin{Bmatrix} 1 \\ 33 \end{Bmatrix} = -re^{-\lambda}\sin^2\theta, \quad \begin{Bmatrix} 1 \\ 44 \end{Bmatrix} = \frac{1}{2}e^{(\nu-\lambda)}\nu', \\
\begin{Bmatrix} 2 \\ 12 \end{Bmatrix} = \begin{Bmatrix} 2 \\ 21 \end{Bmatrix} = \begin{Bmatrix} 3 \\ 13 \end{Bmatrix} = \begin{Bmatrix} 3 \\ 31 \end{Bmatrix} = \frac{1}{r}, \qquad \begin{Bmatrix} 2 \\ 33 \end{Bmatrix} = -\sin\theta\cos\theta, \\
\begin{Bmatrix} 3 \\ 23 \end{Bmatrix} = \begin{Bmatrix} 3 \\ 32 \end{Bmatrix} = \cot\theta, \qquad \begin{Bmatrix} 4 \\ 14 \end{Bmatrix} = \begin{Bmatrix} 4 \\ 41 \end{Bmatrix} = \frac{1}{2}\nu'.
\end{gathered} \tag{47.4}$$

Then from (44.6) and (44.10) the nonzero covariant components of the Ricci tensor **S** are

$$
\begin{aligned}
S_{11} &= \frac{1}{2}\nu'' - \frac{1}{4}\lambda'\nu' + \frac{1}{4}\nu'^2 - \frac{\lambda'}{r},\\
S_{22} &= e^{-\lambda}\left[1 + \frac{1}{2}r(\nu' - \lambda')\right] - 1,\\
S_{33} &= e^{-\lambda}\sin^2\theta\left[1 + \frac{1}{2}r(\nu' - \lambda')\right] - \sin^2\theta.\\
S_{44} &= e^{(\nu-\lambda)}\left[-\frac{\nu''}{2} + \frac{\lambda'\nu'}{4} - \frac{\nu'^2}{4} - \frac{\nu'}{r}\right].
\end{aligned}
\tag{47.5}
$$

From (47.1) we see that the right-hand sides of the four equa ons in (47.5) must all vanish.

From $S_{11} = 0$ and $S_{44} = 0$, we obtain $\lambda' = -\nu'$. Assuming that, as $r \to \infty$, the metric approaches the flat metric $\tilde{\sigma}$ with components

$$\tilde{\Sigma}_{11} = 1, \qquad \tilde{\Sigma}_{22} = r^2, \qquad \tilde{\Sigma}_{33} = r^2\sin^2\theta, \qquad \tilde{\Sigma}_{44} = -1, \tag{47.6}$$

we get $\lambda = -\nu$. Now using this condition in either $S_{22} = 0$ or $S_{33} = 0$, we obtain

$$e^{\nu}(1 + r\nu') = \frac{d}{dr}(re^{\nu}) = 1. \tag{47.7}$$

The preceding equation may be integrated easily, and the solution is

$$e^{\nu} = 1 - \frac{2km}{c^2 r} = e^{-\lambda}, \tag{47.8}$$

where the constant of integration is determined by the condition that $\zeta = -km/r$, the Newtonian potential, satisfies (46.8).

Note. When $\lambda = -\nu$, the conditions $S_{11} = S_{44} = 0$ have an integral $e^{\nu}(1 + r\nu') =$ constant. Hence the solution (47.8) automatically satisfies the requirements $S_{11} = S_{44} = 0$.

We now use the Schwarzschild solution to analyze the geodesics in the neighborhood of the center $r = 0$. As we have remarked before, a timelike geodesic corresponds to the world line of an infinitesimal particle (i.e., whose own gravitational field is neglected), while a signal-like geodesic corresponds to a path of light (i.e., the history of a light front) passing by the center $r = 0$.

We consider first the orbital problem. From (44.15) any timelike geodesic $(x^\alpha(s)) = (r(s), \theta(s), \varphi(s), ct(s))$ satisfies the system of equations

$$\ddot{x}^\alpha + \begin{Bmatrix} \alpha \\ \beta\gamma \end{Bmatrix} \dot{x}^\beta \dot{x}^\gamma = 0, \qquad \alpha = 1, 2, 3, 4, \tag{47.9}$$

and the condition of normalization

$$\sigma(\dot{\mathbf{w}}, \dot{\mathbf{w}}) = \Sigma_{\alpha\beta} \dot{x}^\alpha \dot{x}^\beta = -c^2. \tag{47.10}$$

We set $\alpha = 2$ in (47.9) first. From (47.4)

$$\ddot{\theta} + \frac{2}{r} \dot{r}\dot{\theta} - \sin\theta \cos\theta \dot{\varphi}^2 = 0. \tag{47.11}$$

We consider the special case that the geodesic stays on the plane such that

$$\theta = \frac{\pi}{2}. \tag{47.12}$$

Clearly the preceding assumption is consistent with the equation (47.11). For this special case the remaining three equations in the system (47.9) are

$$\begin{gathered} \ddot{r} - \frac{km/c^2r^2}{1 - 2km/c^2r} \dot{r}^2 - r\left(1 - \frac{2km}{c^2r}\right)\dot{\varphi}^2 + \frac{km}{c^2r^2}\left(1 - \frac{2km}{c^2r}\right)c^2\dot{t}^2 = 0, \\ \ddot{\varphi} + \frac{2}{r} \dot{r}\dot{\varphi} = 0, \qquad (47.13) \\ c\ddot{t} + \frac{2km/c^2r^2}{1 - 2km/c^2r} \dot{r}c\dot{t} = 0. \end{gathered}$$

Since the tangent vector of the world line satisfies the condition (47.10), we have an integral

$$\frac{\dot{r}^2}{1 - 2km/c^2r} + r^2\dot{\varphi}^2 - c^2\dot{t}^2\left(1 - \frac{2km}{c^2r}\right) = -c^2. \tag{47.14}$$

Next, the last two equations of (47.13) give rise to the integrals

$$r^2\dot{\varphi} = h, \tag{47.15a}$$

$$\left(1 - \frac{2km}{c^2r}\right)\dot{t} = q, \tag{47.15b}$$

where h and q are two constants of integration. Substituting (47.15) into

(47.14), we get

$$\dot{r}^2 + r^2\dot{\varphi}^2 - \frac{2km}{r} = c^2q^2 - c^2 + \frac{2kmr}{c^2}\dot{\varphi}^2. \tag{47.16}$$

The integrals (47.15a) and (47.16) may be compared with corresponding results in the Newtonian theory, viz.,

$$\left(\frac{dr}{dt}\right)^2 + r^2\left(\frac{d\varphi}{dt}\right)^2 - \frac{2km}{r} = \frac{2\bar{E}}{\bar{m}}, \tag{47.17a}$$

$$r^2\frac{d\varphi}{dt} = \frac{\bar{H}}{\bar{m}}, \tag{47.17b}$$

where $\bar{m}$, $\bar{E}$, and $\bar{H}$ denote the mass, the total energy, and the moment of momentum relative to the center $r = 0$, respectively, of the orbiting infinitesimal particle, and where m denotes the mass of the particle at the center $r = 0$. It is important that the mass $\bar{m}$ be infinitesimal compared with the mass m, since the problem of two bodies has not been solved in the general theory of relativity. (The Schwarzschild solution has singularities near the center $r = 0$ only. The classical trick of using the center of mass of two particles to solve the problem of two bodies does not work in the relativistic formulation.)

We notice that the main difference between (47.16) and (47.17) is the extra term $(2kmr/c^2)\dot{\varphi}^2$ in the former equation. Also, the independent variable s in the relativistic equations is the proper time, while that in the classical equations is the Newtonian time. A more direct comparison may be made on the orbits of the infinitesimal particle. We eliminate the variable s in (47.16) by using the integral (47.15a), obtaining

$$\left(\frac{1}{r^2}\frac{dr}{d\varphi}\right)^2 + \frac{1}{r^2} = \frac{2km}{h^2r} + \frac{c^2q^2 - c^2}{h^2} + \frac{2km}{c^2r^3}. \tag{47.18}$$

As remarked in the classical formulation, the equation (47.17a) may be simplified by introducing the new variable $u = 1/r$. Then (47.18) becomes

$$\left(\frac{du}{d\varphi}\right)^2 + u^2 = \frac{2km}{h^2}u + \frac{c^2q^2 - c^2}{h^2} + \frac{2km}{c^2}u^3. \tag{47.19}$$

Differentiating the preceding equation with respect to φ and then removing the nonzero common factor $2du/d\varphi$, we obtain

$$\frac{d^2u}{d\varphi^2} + u = \frac{km}{h^2} + \frac{3km}{c^2}u^2. \tag{47.20}$$

In the classical theory the same procedure leads to

$$\frac{d^2u}{d\varphi^2} + u = \frac{km}{h^2}, \tag{47.21}$$

provided that $\bar{H}/\bar{m} = h$. Comparing (47.20) with (47.21), we see that the relativistic correction term $(3km/c^2)u^2$ is extremely small. Indeed, its ratio with the classical term km/h^2 is

$$\frac{3kmu^2}{c^2}\bigg/\frac{km}{h^2} = 3\left(\frac{r\dot{\varphi}}{c}\right)^2, \tag{47.22}$$

where $r\dot{\varphi}$ is of the order of the orbital speed.

The equation (47.20) may be viewed as a nonlinear vibration problem in Newtonian mechanics with φ corresponding to the time and $\frac{1}{2}u^2 - (km/h^2)u - (km/c^2)u^3$ corresponding to the potential energy. Without the third-order term the solution is given by a simple harmonic motion with neutral position at u equal to km/h^2 and the period in φ equal to 2π. This classical solution gives rise to a closed elliptic orbit for the particle. When the third-order term is included in the potential energy function, the neutral position is shifted to the root of the algebraic equation

$$u - \frac{km}{h^2} - \frac{3km}{c^2}u^2 = 0 \tag{47.23}$$

near the value km/h^2, and the period in φ is lengthened to $2\pi + \delta$. Then the orbit may or may not be closed, and the perihelion advances by an angle δ per cycle in u. This correction is small in each cycle, but it will accumulate and become observable after many cycles.

To find an approximate value for the angle δ, we use the following iterative procedure: First, we solve the classical equation (47.21) without the third-order term. The result is

$$u = \frac{km}{h^2}[1 + \varepsilon\cos(\varphi - \varphi_0)], \tag{47.24}$$

where ε and φ_0 are two constants of integration representing the eccentricity and the phase angle of the perihelion of the closed elliptic orbit. Substituting (47.24) into the right-hand side of (47.20), we get an approximation of the nonlinear equation by a linear equation with forcing terms depending on the independent variable φ, viz.,

$$\frac{d^2u}{d\varphi^2} + u = \frac{km}{h^2} + \frac{3k^3m^3}{c^2h^4}\left(1 + \frac{\varepsilon^2}{2}\right) + \frac{6k^3m^3}{c^2h^4}\,\varepsilon\cos(\varphi - \varphi_0) + \frac{3k^3m^3}{2c^2h^4}\,\varepsilon^2\cos 2(\varphi - \varphi_0). \tag{47.25}$$

We recognize immediately that there is a forcing term which is proportional to the base solution given by (47.24). In the theory of vibrations it is known that such a forcing term gives rise to a resonance, and the solution is of the form

$$\tilde{u} = \frac{3k^3m^3}{c^2h^4}\,\varepsilon\varphi\sin(\varphi - \varphi_0), \tag{47.26}$$

where the resonant amplitude is proportional to the independent variable φ. Adding this correction term to the base solution, we get the approximate solution of the nonlinear equation,

$$u = \frac{km}{h^2}\,[1 + \varepsilon\cos(\varphi - \varphi_0 - \varphi\delta)], \tag{47.27}$$

where

$$\delta = \frac{3k^2m^2}{c^2h^2}. \tag{47.28}$$

For the planet Mercury the preceding formula predicts that the accumulated advances of the perihelion in a century amounts to about 43 sec. This result agrees quite well with the observed value in astronomy and is generally regarded as strong evidence in support of the validity of Einstein's field equations.

The problem of the deflection of light near the center $r = 0$ of the Schwarzschild solution may be solved in a similar way. The only change from the previous problem is that the path of light corresponds to a signal-like geodesic whose tangent vector is a signal vector. Hence $\dot{\mathbf{w}}$ satisfies, instead of (47.10),

$$\sigma(\dot{\mathbf{w}}, \dot{\mathbf{w}}) = \Sigma_{\alpha\beta}\dot{x}^\alpha\dot{x}^\beta = 0. \tag{47.29}$$

As a result, the parameter s is unique to within an arbitrary positive factor only. Then (47.14) is replaced by

$$\frac{\dot{r}^2}{1 - 2km/c^2r} + r^2\dot{\varphi}^2 - c^2\dot{t}^2(1 - 2km/c^2r) = 0, \tag{47.30}$$

while (47.15) still holds. By following the procedure as before, we obtain

$$\frac{d^2u}{d\varphi^2} + u = \frac{3km}{c^2} u^2. \tag{47.31}$$

The preceding equation may be compared to the classical equation

$$\frac{d^2u}{d\varphi^2} + u = 0, \tag{47.32}$$

which has a general solution of the form

$$u = u_0 \cos(\varphi - \varphi_0), \tag{47.33}$$

where $r_0 = 1/u_0$ denotes the radius, and where φ_0 denotes the phase angle of the perihelion. The orbit is, of course, a straight line in the (r, φ) plane.

The angular difference between any two consecutive zeros of (47.33) is precisely equal to π. That difference corresponds to the fact that the two directions in which the light path goes toward infinity differ by an angle π. When the nonlinear term is included, the solution of (47.31) is no longer given by (47.33). The angular difference of two consecutive zeros of the solution $u = u(\varphi)$ still has the meaning as before. However, the value is $\pi + \xi$, where ξ is a small positive angle of deflection, which may be estimated by the iterative procedure.

Specifically, we substitute the base solution given by (47.33) into the right-hand side of (47.31), obtaining

$$\frac{d^2u}{d\varphi^2} + u = \frac{3kmu_0^2}{c^2} \cos^2(\varphi - \varphi_0). \tag{47.34}$$

A particular solution of the preceding equation is

$$\tilde{u} = \frac{kmu_0^2}{c^2} [\cos^2(\varphi - \varphi_0) + 2\sin^2(\varphi - \varphi_0)]. \tag{47.35}$$

Then the approximate solution of the nonlinear equation (47.31) with the same radius and the same phase angle as the base solution at the perihelion is

$$u = u_0 \cos(\varphi - \varphi_0) + \frac{kmu_0^2}{c^2} [\cos^2(\varphi - \varphi_0) + 2\sin^2(\varphi - \varphi_0)]. \tag{47.36}$$

Changing the coordinates (r, φ) into the Cartesian coordinates (x, y) by the transformation

$$x = r\cos(\varphi - \varphi_0), \qquad y = r\sin(\varphi - \varphi_0), \tag{47.37}$$

we can rewrite the solution (47.36) as

$$x = r_0 - \frac{km}{c^2 r_0} \frac{x^2 + 2y^2}{(x^2 + y^2)^{1/2}}. \tag{47.38}$$

Hence the equations of the asymptotes are

$$x = r_0 - \frac{km}{c^2 r_0} (\pm 2y). \tag{47.39}$$

Thus the deflection angle ξ is given approximately by

$$\xi = \frac{4km}{c^2 r_0}. \tag{47.40}$$

For the gravitational field of the sun at a distance equal to the sun's radius the preceding formula predicts an angle of deflection amounting to about 1.75 sec. This value also agrees quite well with the observed values obtained at several total solar eclipses.

Before closing this section, we remark that the Schwarzschild solution becomes singular not only at $r = 0$ but also at $r = 2km/c^2$; cf. (47.8). The latter singularity, which has no counterpart in the classical theory at all, is known as the *Schwarzschild singularity*. This peculiar singularity cannot be observed, however, since mass cannot be compacted into such high density as to make it outside the material medium. Indeed, if $m = 1$ g, then the radius of the Schwarzschild singularity is only

$$r = \frac{2k}{c^2} \times 1 = 1.48 \times 10^{-28} \text{ cm}. \tag{47.41}$$

If m is of the order of the solar mass, $m = 1.99 \times 10^{33}$ g, the radius of the singularity is of the order

$$r = 2.95 \times 10^5 \text{ cm}, \tag{47.42}$$

which is much smaller than the radius, 6.95×10^{10} cm, of the sun.

48. The Action Principle

In analytical mechanics we have shown that the trajectories of a conservative system may be characterized either by the system of Lagrange's equations or by Hamilton's principle. The former is a system of differential equations, and the latter is a variational principle. Trajectories of the system are, on the one hand, solution curves of Lagrange's equations and, on the other hand, extremal curves of the variational integral of the Lagrangian function taken over a class of curves satisfying certain boundary conditions and smoothness conditions. These results have been discussed in detail in Sections 5 and 7.

We note that the system of differential equations is essentially determined by the trajectories. Indeed, if the trajectories are given by a parametrized family of coordinate functions, then the differential equations may be obtained by eliminating the parameters from the coordinate functions and their time derivatives. The variational integral, however, is not determined by the trajectories. Clearly, a function cannot be determined by its extremal points, and the trajectories are merely the extremal curves of the variational integral over a family of curves. Hamilton's principle gives but one variational integral, whose extremal curves coincide with the trajectories. Many other variational integrals possessing the same extremal curves exist.

In much the same way a gravitational potential may be characterized either by some differential equations or by some variational principles. We have shown in Section 38 that the Newtonian potential ζ satisfies the Laplace equation

$$\nabla^2\zeta = 0 \tag{48.1}$$

in a vacuum and the Poisson equation

$$\nabla^2\zeta = 4\pi k\varrho \tag{48.2}$$

in a material medium. Now we show that ζ may be characterized by several variational principles also.

Let $\mathscr{D}$ be a bounded domain in an instantaneous space in Newtonian space–time. We define the Dirichlet integral $D(\tilde{\zeta})$ for smooth functions $\tilde{\zeta}$ on the closure $\bar{\mathscr{D}}$ by

$$D(\tilde{\zeta}) = \int_{\mathscr{D}} \operatorname{grad}\tilde{\zeta}\cdot\operatorname{grad}\tilde{\zeta}\, dx. \tag{48.3}$$

Then we have the following variational principle: ζ is a Newtonian potential in $\mathscr{D}$ for vacuum if and only if it is an extremum for the Dirichlet integral among all smooth functions $\tilde{\zeta}$ having the same boundary values as ζ.

This variational principle may be proved easily in the following way: We take the variation of the Dirichlet integral D at a particular function

ζ over the set of functions having the same boundary values as ζ:

$$\begin{aligned}\delta D(\zeta) &= \int_{\mathscr{D}} 2 \operatorname{grad} \zeta \cdot \operatorname{grad} \delta\zeta \, dx \\ &= \int_{\mathscr{D}} 2 \operatorname{div}(\delta\zeta \operatorname{grad} \zeta) \, dx - \int_{\mathscr{D}} 2\delta\zeta \, \nabla^2\zeta \, dx \\ &= \int_{\partial\mathscr{D}} 2\delta\zeta \operatorname{grad} \zeta \cdot \mathbf{n} \, d\sigma - \int_{\mathscr{D}} 2\delta\zeta \, \nabla^2\zeta \, dx \\ &= \int_{\mathscr{D}} (-2\nabla^2\zeta) \, \delta\zeta \, dx, \end{aligned} \tag{48.4}$$

where the surface integral vanishes since $\delta\zeta = 0$ on the boundary. Consequently, $\delta D(\zeta) = 0$ for all $\delta\zeta$ if and only if ζ satisfies (48.1). Thus the variational principle is proved.

Note. In (48.4) $\delta\zeta$ denotes the variation of ζ and $\delta D(\zeta)$ denotes the variation of D at ζ. These notions are standard in the calculus of variations. We may use a 1-parameter family of functions to express the variations as shown in Section 7. Specifically, we consider the family of functions

$$\zeta_\varepsilon = \zeta_\varepsilon(\mathbf{x}) = \zeta(\mathbf{x}) + \varepsilon\eta(\mathbf{x}), \qquad \mathbf{x} \in \mathscr{D}, \tag{48.5}$$

where ε is a parameter, and where $\eta(\mathbf{x})$ is a smooth function on $\bar{\mathscr{D}}$ satisfying the boundary condition

$$\eta(\mathbf{x}) = 0, \qquad x \in \partial\mathscr{D}. \tag{48.6}$$

When $\varepsilon = 0$, the function ζ_0 reduces to the function ζ. Thus ζ_ε corresponds to a 1-parameter family of variations from the function ζ in the direction of the function η. Using the 1-parameter family ζ_ε, we can rewrite the equation (48.4) as

$$\frac{d}{d\varepsilon} D(\zeta_\varepsilon) \big|_{\varepsilon=0} = \int_{\mathscr{D}} \left[-2\nabla^2\zeta_\varepsilon \frac{d\zeta_\varepsilon}{d\varepsilon} \right]_{\varepsilon=0} dx = \int_{\mathscr{D}} (-2\nabla^2\zeta)\eta \, dx. \tag{48.7}$$

Since the notations $\delta\zeta$ and δD are more compact, we shall use them throughout this section. All equations expressed in terms of these notations may be rewritten in terms of the derivatives with respect to the parameter ε if we use the 1-parameter family ζ_ε; we have just shown that the equation (48.4) may be rewritten as the equation (48.7).

The preceding variational principle asserts that $D(\zeta)$ is an extremum at ζ over the class of functions ξ having the same boundary values as ζ. In fact $D(\zeta)$ is a minimum among the values $D(\xi)$, since from (48.4)

$$\begin{aligned} D(\zeta+\delta\zeta) - D(\zeta) &= \int_{\mathscr{D}} [\operatorname{grad}(\zeta+\delta\varrho)\cdot\operatorname{grad}(\zeta+\delta\zeta) - \operatorname{grad}\zeta\cdot\operatorname{grad}\zeta]\,dx \\ &= \int_{\mathscr{D}} 2\operatorname{grad}\zeta\cdot\operatorname{grad}\delta\zeta\,dx + \int_{\mathscr{D}} \operatorname{grad}\delta\zeta\cdot\operatorname{grad}\delta\zeta\,dx \\ &= \int_{\mathscr{D}} \operatorname{grad}\delta\zeta\cdot\operatorname{grad}\delta\zeta\,dx, \end{aligned} \tag{48.8}$$

which is positive unless $\delta\zeta = \text{const} = 0$, where we have used the boundary condition on $\delta\zeta$ to determine the value of the constant. An extremum of a variational integral in general need not be a minimum, of course. The fact that $D(\zeta)$ is actually a minimum is asserted by the Dirichlet principle. In this section we are mainly interested in various variational principles for the gravitational potential, both in the classical theory and in the relativistic theory.

We call the coefficient of $\delta\zeta$ in the integrand of the variation $\delta D(\zeta)$ the *variational derivative* of D at ζ; i.e.,

$$\delta D(\zeta) = \int_{\mathscr{D}} \frac{\delta D(\zeta)}{\delta\zeta}\,\delta\zeta\,dx, \qquad \frac{\delta D(\zeta)}{\delta\zeta} \equiv -2\nabla^2\zeta \tag{48.9}$$

Using this concept, we can rewrite the field equation (48.1) as

$$\frac{\delta D(\zeta)}{\delta\zeta} = 0, \tag{48.10}$$

and the field equation (48.2) as

$$\frac{\delta D(\zeta)}{\delta\zeta} = -2\pi k\varrho, \tag{48.11}$$

or, equivalently,

$$\varrho = -\frac{1}{2\pi k}\,\frac{\delta D(\zeta)}{\delta\zeta}, \tag{48.12}$$

which asserts that the mass density is determined by the gravitational potential ζ through the variational derivative of the Dirichlet integral.

The Dirichlet integral is just one example of variational integrals whose extremum over a class of functions ξ coincides with the Newtonian gravitational potential ζ of a vacuum. Another example is the integral

$$A(\tilde{\zeta}) = \int_{\mathscr{D}} -(\tilde{\zeta}\nabla^2\tilde{\zeta} + \tfrac{1}{2}\operatorname{grad}\tilde{\zeta}\cdot\operatorname{grad}\tilde{\zeta})\,dx. \tag{48.13}$$

We claim that ζ is a solution of the field equation (48.1) in the domain $\mathscr{D}$ if and only if it is an extremum of the variational integral $A(\tilde{\zeta})$ over the class of smooth functions $\tilde{\zeta}$ such that the boundary values of $\tilde{\zeta}$ and grad $\tilde{\zeta}$ coincide with those of ζ and grad ζ, respectively.

This variational principle may be proved easily as before. We take the variation of A from (48.13)

$$\begin{aligned} \delta A(\zeta) &= \int_{\mathscr{D}} -(\delta\zeta\nabla^2\zeta + \zeta\nabla^2\delta\zeta + \operatorname{grad}\zeta\cdot\operatorname{grad}\delta\zeta)\,dx \\ &= \int_{\mathscr{D}} -\delta\zeta\nabla^2\zeta\,dx - \int_{\mathscr{D}} \operatorname{div}(\zeta\operatorname{grad}\delta\zeta)\,dx \\ &= \int_{\mathscr{D}} -\delta\zeta\nabla^2\zeta\,dx - \int_{\partial\mathscr{D}} \zeta\operatorname{grad}\delta\zeta\,d\sigma \\ &= \int_{\mathscr{D}} (-\nabla^2\zeta)\,\delta\zeta\,dx, \end{aligned} \tag{48.14}$$

where we have used the boundary condition on grad $\delta\zeta$ to eliminate the surface integral. It follows from (48.14) that $\delta A(\zeta) = 0$ for all $\delta\zeta$ if and only if $\nabla^2\zeta = 0$. Thus the variational principle is proved.

Using the concept of variational derivative defined before, we have

$$\frac{\delta A(\zeta)}{\delta\zeta} = -\nabla^2\zeta. \tag{48.15}$$

Thus for a vacuum we can rewrite the field equation (48.1) as

$$\frac{\delta A(\zeta)}{\delta\zeta} = 0, \tag{48.16}$$

while for a material medium we can rewrite the field equation (48.2) as

$$\frac{\delta A(\zeta)}{\delta\zeta} = -4\pi k\varrho, \tag{48.17}$$

or, equivalently,

$$\varrho = -\frac{1}{4\pi k}\,\frac{\delta A(\zeta)}{\delta\zeta}, \tag{48.18}$$

which asserts that the mass density ϱ may be determined by ζ through the variational derivative of A.

Note. Like the Dirichlet integral D, the variational integral A also takes a minimum at ζ among the class of functions $\tilde{\zeta}$. Indeed, from (48.1), (48.13), and (48.14) the difference between $A(\zeta + \delta\zeta)$ and $A(\zeta)$ is given by

$$\begin{aligned} A(\zeta + \delta\zeta) - A(\zeta) &= \int_{\mathscr{D}} -(\delta\zeta\nabla^2\delta\zeta + \tfrac{1}{2}\operatorname{grad}\delta\zeta \cdot \operatorname{grad}\delta\zeta)\,dx \\ &= \int_{\mathscr{D}} (-\operatorname{div}(\delta\zeta \operatorname{grad}\delta\zeta) + \operatorname{grad}\delta\zeta \cdot \operatorname{grad}\delta\zeta)\,dx \\ &\quad - \int_{\mathscr{D}} \tfrac{1}{2}\operatorname{grad}\delta\zeta \cdot \operatorname{grad}\delta\zeta\,dx \\ &= \int_{\mathscr{D}} \tfrac{1}{2}\operatorname{grad}\delta\zeta \cdot \operatorname{grad}\delta\zeta\,dx, \end{aligned} \tag{48.19}$$

which is positive unless $\delta\zeta = \text{const} = 0$, where we have used the boundary condition on $\delta\zeta$ to determine the value of the constant.

Now in the general theory of relativity the Minkowskian metric plays the role of the gravitational potential. The system of field equations is

$$G_{\alpha\beta} = 0 \tag{48.20}$$

for a vacuum and is

$$G_{\alpha\beta} = -\frac{8\pi k}{c^4}\,\Theta_{\alpha\beta} \tag{48.21}$$

for a material medium. Notice that the left-hand side of (48.20) and (48.21) are formed by partial derivatives up to the second order in the components $\Sigma_{\alpha\beta}$ of the Minkowskian metric σ. Thus the systems of field equations (48.20) and (48.21) are comparable to the field equations (48.1) and (48.2), respectively.

It turns out that $G_{\alpha\beta}$ may be determined by the variational derivative of the integral

$$A(\tilde{\sigma}) = \int_{\mathscr{D}} \tilde{S}\tilde{\Xi} = \int_{\mathscr{D}} \tilde{S}(-\tilde{\Sigma})^{1/2}\,dw, \tag{48.22}$$

where $\tilde{S}$ and $\tilde{\Xi}$ denote the curvature scalar and the unit density tensor based on the Minkowskian metric $\tilde{\sigma}$, which is required to satisfy the boundary conditions $\tilde{\Sigma}_{\alpha\beta} = \Sigma_{\alpha\beta}$ and $\partial\tilde{\Sigma}_{\alpha\beta}/\partial x^\gamma = \partial\Sigma_{\alpha\beta}/\partial x^\gamma$ on $\partial\mathscr{D}$. Of course the domain $\mathscr{D}$ here is contained in the 4-dimensional differentiable manifold $\mathscr{E}$, (x^γ) is any coordinate system in $\mathscr{D}$, and $dw = dx^1\, dx^2\, dx^3\, dx^4$.

Note. It can be verified easily from the transformation law of the components of the metric that if the boundary conditions $\tilde{\Sigma}_{\alpha\beta} = \Sigma_{\alpha\beta}$ and $\partial\tilde{\Sigma}_{\alpha\beta}/\partial x^\gamma = \partial\Sigma_\alpha{}^\gamma/\partial x^\gamma$ are satisfied on $\partial\mathscr{D}$ relative to any one coordinate system (x^γ) in $\mathscr{D}$, then the same are satisfied relative to all other coordinate systems in $\mathscr{D}$. Thus the boundary conditions are actually conditions on the metrics $\tilde{\sigma}$ and σ independent of the choice of the coordinate system (x^γ).

Now we claim that

$$\frac{\delta A(\sigma)}{\delta\Sigma^{\alpha\beta}} = G_{\alpha\beta}. \tag{48.23}$$

Hence for a vacuum we can rewrite the field equations (48.20) as

$$\frac{\delta A(\sigma)}{\delta\Sigma^{\alpha\beta}} = 0, \tag{48.24}$$

while for a material medium we can rewrite the field equations (48.21) as

$$\frac{\delta A(\sigma)}{\delta\Sigma^{\alpha\beta}} = -\frac{8\pi k}{c^4}\,\Theta_{\alpha\beta}. \tag{48.25}$$

The condition (48.24) means that the Minkowskian metric σ in a vacuum is an extremum of the variational integral A over the class of Minkowskian metrics $\tilde{\sigma}$ having the same boundary values and the same first derivatives on the boundary as σ. For a vacuum the assertion that

$$\delta A(\sigma) = \int_{\mathscr{D}} \frac{\delta A(\sigma)}{\delta\Sigma^{\alpha\beta}}\,\delta\Sigma^{\alpha\beta}\,dw = 0 \tag{48.26}$$

for all variations $\delta\Sigma^{\alpha\beta}$ of the Minkowskian metric satisfying the aforementioned boundary conditions is known as the *action principle*.

Note. The invariant integral $A(\sigma)$ is, in fact, proportional to the usual notion of an action integral. Indeed, from (44.12) the curvature scalar S is related to the trace G of the Einstein tensor $\mathbf{G}$ by

$$G = -S. \tag{48.27}$$

From (48.21) G is related to the trace Θ of the stress-energy-momentum tensor $\mathbf{\Theta}$ by

$$G = -\frac{8\pi k}{c^4}\Theta. \tag{48.28}$$

Now recall that the trace of $\mathbf{\Theta}$ is dominated by the intrinsic energy $-\varrho c^2$. It follows that $A(\sigma)$ is given approximately by

$$A(\sigma) = \int_{\mathscr{D}} -\frac{8\pi k}{c^2}\varrho(-\Sigma)^{1/2}\,dw, \tag{48.29}$$

which is proportional to the usual definition of an action, i.e., an integral of energy with respect to time. In particular, $A(\sigma) = 0$ in a vacuum.

To prove the formula (48.23), we calculate the variation $\delta A(\sigma)$ directly from (48.22):

$$\delta A(\sigma) = \int_{\mathscr{D}} \{S_{\alpha\beta}\,\delta[\Sigma^{\alpha\beta}(-\Sigma)^{1/2}] + \Sigma^{\alpha\beta}(-\Sigma)^{1/2}\,\delta S_{\alpha\beta}\}\,dw, \tag{48.30}$$

where the variation $\delta[\Sigma^{\alpha\beta}(-\Sigma)^{1/2}]$ is given as usual by

$$\delta[\Sigma^{\alpha\beta}(-\Sigma)^{1/2}] = (-\Sigma)^{1/2}(\delta\Sigma^{\alpha\beta} - \tfrac{1}{2}\Sigma^{\alpha\beta}\Sigma_{\gamma\delta}\delta\Sigma^{\gamma\delta}). \tag{48.31}$$

Substituting this result into the first term of the integrand and using (44.12), we get

$$S_{\alpha\beta}\delta[\Sigma^{\alpha\beta}(-\Sigma)^{1/2}] = G_{\alpha\beta}(-\Sigma)^{1/2}\delta\Sigma^{\alpha\beta}. \tag{48.32}$$

Hence to prove (48.23) we must show that the integral involving the variation $\delta S_{\alpha\beta}$ reduces to a boundary term.

From (44.6) and (44.10)

$$\delta S_{\alpha\beta} = \frac{\partial}{\partial x^\beta}\delta\begin{Bmatrix}\gamma\\ \alpha\gamma\end{Bmatrix} - \frac{\partial}{\partial x^\gamma}\delta\begin{Bmatrix}\gamma\\ \alpha\beta\end{Bmatrix} + \begin{Bmatrix}\lambda\\ \alpha\gamma\end{Bmatrix}\delta\begin{Bmatrix}\gamma\\ \lambda\beta\end{Bmatrix} + \begin{Bmatrix}\gamma\\ \lambda\beta\end{Bmatrix}\delta\begin{Bmatrix}\lambda\\ \alpha\gamma\end{Bmatrix}$$
$$- \begin{Bmatrix}\lambda\\ \alpha\beta\end{Bmatrix}\delta\begin{Bmatrix}\gamma\\ \lambda\gamma\end{Bmatrix} - \begin{Bmatrix}\gamma\\ \lambda\gamma\end{Bmatrix}\delta\begin{Bmatrix}\lambda\\ \alpha\beta\end{Bmatrix}. \tag{48.33}$$

Now recall that the Christoffel symbols satisfy the set of transformation rules:

$$\overline{\begin{Bmatrix}\alpha\\ \beta\gamma\end{Bmatrix}} = \begin{Bmatrix}\varphi\\ \mu\nu\end{Bmatrix}\frac{\partial\bar{x}^\alpha}{\partial x^\varphi}\frac{\partial x^\mu}{\partial\bar{x}^\beta}\frac{\partial x^\nu}{\partial\bar{x}^\gamma} - \frac{\partial^2 x^\nu}{\partial\bar{x}^\beta\,\partial\bar{x}^\gamma}\frac{\partial\bar{x}^\alpha}{\partial x^\nu} \tag{48.34}$$

under a change of coordinate system from (x^α) to $(\bar{x}^\alpha)$; cf. equation (56.15) in Section 56, IVT-2. Hence $\delta\{{}^{\alpha}_{\beta\gamma}\}$ are the components of a third-order tensor field on $\mathscr{D}$. By using the formula for the covariant derivative of a third-order tensor field, we can rewrite (48.33) as

$$\delta S_{\alpha\beta} = \left(\delta\begin{Bmatrix}\gamma \\ \alpha\gamma\end{Bmatrix}\right)_{,\beta} - \left(\delta\begin{Bmatrix}\gamma \\ \alpha\beta\end{Bmatrix}\right)_{,\gamma}. \tag{48.35}$$

Substituting this result into the second term of the integrand in (48.30), we get

$$\Sigma^{\alpha\beta}(-\Sigma)^{1/2}\delta S_{\alpha\beta} = (-\Sigma)^{1/2}\left(\Sigma^{\alpha\beta}\delta\begin{Bmatrix}\gamma \\ \alpha\gamma\end{Bmatrix} - \Sigma^{\gamma\lambda}\delta\begin{Bmatrix}\beta \\ \gamma\lambda\end{Bmatrix}\right)_{,\beta}. \tag{48.36}$$

Now using the identity

$$(-\Sigma)^{1/2}(V^{\beta})_{,\beta} = \frac{\partial}{\partial x^{\beta}}\,[(-\Sigma)^{1/2}V^{\beta}] \tag{48.37}$$

for any vector field $\mathbf{V}$, we see that the contribution of the second term in the integrand of (48.30) may be expressed as a surface integral which vanishes by virtue of the boundary condition on the first derivatives of $\tilde{\Sigma}_{\alpha\beta}$; i.e., $\delta\{{}^{\alpha}_{\beta\gamma}\} = 0$ on $\partial\mathscr{D}$. Thus

$$\delta A(\sigma) = \int_{\mathscr{D}} G_{\alpha\beta}\delta\Sigma^{\alpha\beta}(-\Sigma)^{1/2}\,dw = \int_{\mathscr{D}} (G_{\alpha\beta})\delta\Sigma^{\alpha\beta}\mathbf{\Sigma}, \tag{48.38}$$

which implies (48.23) by the definition of the variational derivative.

Note. To get Einstein's field equations including the cosmologic constant C, we simply change the definition of the integral $A(\tilde{\sigma})$ to

$$A(\tilde{\sigma}) = \int_{\mathscr{D}} (\tilde{S} + 2C)\tilde{\mathbf{\Xi}} \tag{48.39}$$

Then (48.23) is replaced by

$$\frac{\delta A(\sigma)}{\delta\Sigma^{\alpha\beta}} = G_{\alpha\beta} - C\Sigma_{\alpha\beta}. \tag{48.40}$$

In other words the metric $\Sigma_{\alpha\beta}$ is -2 times the variational derivative of the volume integral; i.e.,

$$\frac{\delta V(\sigma)}{\delta\Sigma^{\alpha\beta}} = -\tfrac{1}{2}\Sigma_{\alpha\beta}, \tag{48.41}$$

where $V(\tilde{\sigma})$ is defined by

$$V(\tilde{\sigma}) = \int_{\mathscr{D}} \tilde{\mathbf{\Xi}}. \tag{48.42}$$

The formula (48.41) follows directly from the variation of the volume integral V:

$$\delta V(\sigma) = \int_{\mathscr{D}} \delta(-\Sigma)^{1/2}\, dw$$
$$= \int_{\mathscr{D}} -\tfrac{1}{2}\Sigma_{\alpha\beta}\delta\Sigma^{\alpha\beta}(-\Sigma)^{1/2}\, dw$$
$$= \int_{\mathscr{D}} (-\tfrac{1}{2}\Sigma_{\alpha\beta})\delta\Sigma^{\alpha\beta}\tilde{\Xi}. \tag{48.43}$$

49. Action and Coaction

In the preceding section we explained the relation between the Einstein tensor **G** of the Minkowskian metric σ and the variational derivative $\delta A(\sigma)/d\sigma$ of the action integral A, which is an invariant integral of a certain 4-form over a domain in the 4-dimensional differentiable manifold $\mathscr{E}$. The variational derivative of A is taken with respect to a class of Minkowskian metrics $\tilde{\sigma}$ satisfying certain boundary conditions. For a vacuum we have shown that the metric σ is an extremum of A among the metrics $\tilde{\sigma}$. In this section we explore another relation between the metric σ and 4-forms on $\mathscr{E}$. Instead of considering the special status of the metric σ among the metrics $\tilde{\sigma}$ relative to a certain 4-form, we now hold the metric σ fixed and show that it gives rise to an operator on 4-forms. It turns out that, conversely, this operator may be used to characterize the metric σ.

Let **A** be any 4-form on $\mathscr{E}$. We call **A** an *action density*, and we define the *action* $A[\mathscr{D}]$ of **A** over an (oriented) 4-dimensional domain $\mathscr{D}$ in $\mathscr{E}$ by

$$A[\mathscr{D}] = \int_{\mathscr{D}} \mathbf{A}. \tag{49.1}$$

The action integral considered in the preceding section is a special case of this concept, such that the 4-form is $S\Xi$. Since S and Ξ are determined by the metric σ and the orientation on $\mathscr{E}$, we regard the action integral $A(\sigma)$ as a function of σ, which belongs to a class of metrics $\tilde{\sigma}$ satisfying certain boundary conditions. Here, the metric σ is held fixed, and we consider the action of a 4-form **A** in general over a 4-dimensional domain $\mathscr{D}$ in $\mathscr{E}$. Thus $A[\mathscr{D}]$ is a function of the pair $(\mathbf{A}, \mathscr{D})$.

Since $\mathscr{E}$ is 4-dimensional, every 4-form in $\mathscr{E}$ is closed; i.e.,

$$d\mathbf{A} = \mathbf{0}. \tag{49.2}$$

Hence by the Poincaré lemma (cf. Section 52, IVT-2) there exists, locally, an *action potential* **P**, which is a 3-form such that

$$\mathbf{A} = d\mathbf{P}. \tag{49.3}$$

As usual the 3-form **P** is unique to within an additive closed 3-form only. We claim that there is a particular action potential **P**, which satisfies the following condition: The dual $\mathbf{D}_3(\mathbf{P})$ of the action potential **P** is a closed 1-form; i.e.,

$$\mathbf{G} = \mathbf{D}_3\mathbf{P}, \tag{49.4a}$$

$$d\mathbf{G} = \mathbf{0}, \tag{49.4b}$$

where the duality operator $\mathbf{D}_3$ is induced at each point $\mathbf{w} \in \mathscr{E}$ by the Minkowskian metric σ and the orientation on the tangent space $\mathscr{E}_{\mathbf{w}}$; cf. (40.27) and (40.32). When we impose the preceding condition on the action potential **P**, the 1-form **G** defined by (49.4a) also has a potential Υ, which is a smooth function (i.e., a 0-form) such that

$$\mathbf{G} = d\Upsilon. \tag{49.5}$$

To prove that an appropriate action potential satisfying (49.4) exists for any given action density **A**, we substitute (49.5) into (49.4) and then into (49.3), obtaining

$$\mathbf{A} = d\mathbf{D}_1\, d\Upsilon, \tag{49.6}$$

where we have used the fact that the inverse of $\mathbf{D}_3$ is just $\mathbf{D}_1$; cf. (40.30) and (40.32). We now apply the duality operator $\mathbf{D}_4$ on (49.6), and the result is a scalar equation (i.e., an equation in 0-forms)

$$M = \mathbf{D}_4\, d\mathbf{D}_1\, d\Upsilon, \tag{49.7}$$

where M denotes the dual of the action density **A**, viz.,

$$M = \mathbf{D}_4\mathbf{A}, \tag{49.8a}$$

$$\mathbf{A} = -M\boldsymbol{\Xi}. \tag{49.8b}$$

Now for a 0-form Υ we define the Laplacian[1] Lap Υ by

$$\operatorname{Lap}\Upsilon \equiv \operatorname{Div}(d\Upsilon), \tag{49.9}$$

where Div denotes the divergence with respect to the Minkowskian metric σ.

[1] For the general concept of the Laplacian of a differential form, see W. V. D. Hodge, *Theory and Applications of Harmonic Integrals*, Cambridge University Press, London and New York, 1952.

Relative to any coordinate system (x^α) the Laplacian is given explicitly by

$$\operatorname{Lap} \Upsilon = \left(\Sigma^{\alpha\beta} \frac{\partial \Upsilon}{\partial x^\alpha}\right)_{,\beta} = \frac{1}{(-\Sigma)^{1/2}} \frac{\partial}{\partial x^\beta}\left((-\Sigma)^{1/2}\Sigma^{\alpha\beta} \frac{\partial \Upsilon}{\partial x^\alpha}\right), \tag{49.10}$$

where the comma denotes the covariant derivative as usual.

Note. For a Minkowskian space–time with a constant metric there are Lorentz systems (x^α) in which the component matrix of the metric is the constant matrix $[\Delta_{\alpha\beta}] = \operatorname{diag}(1, 1, 1, -1)$. Then the right-hand side of (49.10) reduces to

$$\frac{\partial}{\partial x^\beta}\left(\Delta^{\alpha\beta} \frac{\partial \Upsilon}{\partial x^\alpha}\right) = \nabla^2 \Upsilon - \frac{1}{c^2} \frac{\partial^2 \Upsilon}{\partial t^2}. \tag{49.11}$$

The Laplacian in this special case is usually denoted by the symbol $\Box^2$, viz.,

$$\Box^2 \Upsilon = \nabla^2 \Upsilon - \frac{1}{c^2} \frac{\partial^2 \Upsilon}{\partial t^2}. \tag{49.12}$$

The right-hand side of (49.12) is invariant under a Lorentz transformation from one Lorentz system to another Lorentz system but is not invariant under a coordinate transformation in general. For a Minkowskian manifold $\mathscr{E}$ Lorentz systems may or may not exist. Then the Laplacian is defined by (49.9), which has the coordinate form (49.10). The right-hand side of (49.10) is invariant under any coordinate transformation as it should be, since the definition (49.9) does not depend on any coordinate system.

Now we claim that the right-hand side of (49.7) is just the negative of the Laplacian of Υ; i.e.,

$$\operatorname{Lap} \Upsilon = -\mathbf{D}_4\, d\mathbf{D}_1\, d\Upsilon. \tag{49.13}$$

Hence (49.7) is just the generalized Poisson equation

$$\operatorname{Lap} \Upsilon = -M \tag{49.14}$$

on the Minkowskian manifold $\mathscr{E}$.

Note. For a Minkowskian space–time the formula (49.13) may be verified easily by using a Lorentz system (x^α). Indeed, since $d\Upsilon$ has the component form

$$d\Upsilon = \frac{\partial \Upsilon}{\partial x^1}\,\mathbf{e}^1 + \frac{\partial \Upsilon}{\partial x^2}\,\mathbf{e}^2 + \frac{\partial \Upsilon}{\partial x^3}\,\mathbf{e}^3 + \frac{\partial \Upsilon}{\partial x^4}\,\mathbf{e}^4, \tag{49.15}$$

from (40.30) its dual $\mathbf{D}_1\, d\Upsilon$ has the component form

$$\mathbf{D}_1\, d\Upsilon = \frac{\partial \Upsilon}{\partial x^1}\, \mathbf{e}^2 \wedge \mathbf{e}^3 \wedge \mathbf{e}^4 + \frac{\partial \Upsilon}{\partial x^2}\, \mathbf{e}^3 \wedge \mathbf{e}^1 \wedge \mathbf{e}^4 + \frac{\partial \Upsilon}{\partial x^3}\, \mathbf{e}^1 \wedge \mathbf{e}^2 \wedge \mathbf{e}^4 + \frac{\partial \Upsilon}{\partial x^4}\, \mathbf{e}^1 \wedge \mathbf{e}^2 \wedge \mathbf{e}^3. \tag{49.16}$$

Taking the exterior derivative of (49.16), we obtain

$$d\mathbf{D}_1\, d\Upsilon = \left(\frac{\partial^2 \Upsilon}{\partial x^1 \partial x^1} + \frac{\partial^2 \Upsilon}{\partial x^2 \partial x^2} + \frac{\partial^2 \Upsilon}{\partial x^3 \partial x^3} - \frac{\partial^2 \Upsilon}{\partial x^4 \partial x^4}\right) \mathbf{e}^1 \wedge \mathbf{e}^2 \wedge \mathbf{e}^3 \wedge \mathbf{e}^4 = \Box^2 \Upsilon \boldsymbol{\Xi}. \tag{49.17}$$

Then from (44.27) the dual of this 4-form is

$$\mathbf{D}_4\, d\mathbf{D}_1\, d\Upsilon = -\Box^2 \Upsilon. \tag{49.18}$$

Thus the formula (49.13) is proved in this simple case. The preceding proof is not valid for the Minkowskian manifold $\mathscr{E}$, however, since a Lorentz system may not exist.

To verify the formula (49.13) in general, we recall first that the duality operator $\mathbf{D}_r$ is defined by the condition

$$\sigma_{4-r}(\mathbf{D}_r \boldsymbol{\Phi}, \boldsymbol{\Psi}) = \sigma_4(\boldsymbol{\Phi} \wedge \boldsymbol{\Psi}, \boldsymbol{\Xi}) \tag{49.19}$$

for all r-forms $\boldsymbol{\Phi}$ and all $(4 - r)$-forms $\boldsymbol{\Psi}$; cf. (40.27). Using (40.28), we can express the left-hand side of (49.19) in the coordinate form

$$\sigma_{4-r}(\mathbf{D}_r \boldsymbol{\Phi}, \boldsymbol{\Psi}) = \frac{1}{(4-r)!} \Sigma^{\alpha_1 \beta_1} \cdots \Sigma^{\alpha_{4-r} \beta_{4-r}} (D_r \Phi)_{\alpha_1 \cdots \alpha_{4-r}} \Psi_{\beta_1 \cdots \beta_{4-r}} \tag{49.20}$$

and the right-hand side in the coordinate form

$$\sigma_4(\boldsymbol{\Phi} \wedge \boldsymbol{\Psi}, \boldsymbol{\Xi}) = \frac{1}{4!} \Sigma^{\alpha_1 \beta_1} \cdots \Sigma^{\alpha_4 \beta_4} \frac{1}{r!} \frac{1}{(4-r)!} \delta^{\mu_1 \cdots \mu_r \lambda_1 \cdots \lambda_{4-r}}_{\alpha_1 \cdots \qquad \alpha_4} \times \Phi_{\mu_1 \cdots \mu_r} \Psi_{\lambda_1 \cdots \lambda_{4-r}} (-\Sigma)^{1/2} \varepsilon_{\beta_1 \cdots \beta_4}, \tag{49.21}$$

where we have used the fact that the component form of the unit density tensor $\boldsymbol{\Xi}$ is

$$\boldsymbol{\Xi} = (-\Sigma)^{1/2} \varepsilon_{\alpha_1 \cdots \alpha_4}\, dx^{\alpha_1} \otimes \cdots \otimes dx^{\alpha_4} = \frac{(-\Sigma)^{1/2}}{4!} \varepsilon_{\alpha_1 \cdots \alpha_4}\, dx^{\alpha_1} \wedge \cdots \wedge dx^{\alpha_4}. \tag{49.22}$$

As usual $\varepsilon_{\alpha_1 \cdots \alpha_4}$ denotes the skew-symmetric symbol for a 4-dimensional space; cf. Section 20, IVT-1. Now since the condition (49.19) holds for all $(4 - r)$-forms $\mathbf{\Psi}$, combining (49.20) and (49.21) and eliminating $\mathbf{\Psi}$, we get

$$(D_r\Phi)_{\alpha_1\cdots\alpha_{4-r}} = \frac{1}{r!}(-\Sigma)^{1/2}\Sigma^{\beta_1\gamma_1}\cdots\Sigma^{\beta_r\gamma_r}\Phi_{\gamma_1\cdots\gamma_r}\varepsilon_{\beta_1\cdots\beta_r\alpha_1\cdots\alpha_{4-r}} \tag{49.23}$$

The preceding component formula characterizes the duality operator $\mathbf{D}_r$ relative to a positive coordinate system in general. In the derivation of this component formula we have used the formulas (20.13), (21.16), and (38.21) in IVT-1.

Now using the component formula (49.23), we can verify the identity (49.13) by direct calculation. Specifically, the component form of $d\Upsilon$ is

$$d\Upsilon = \frac{\partial\Upsilon}{\partial x^\alpha}dx^\alpha. \tag{49.24}$$

Then from (49.23) with $r = 1$ and $\mathbf{\Phi} = d\Upsilon$

$$\mathbf{D}_1\, d\Upsilon = (-\Sigma)^{1/2}\Sigma^{\alpha\nu}\frac{\partial\Upsilon}{\partial x^\nu}\varepsilon_{\alpha\mu_2\mu_3\mu_4}dx^{\mu_2}\otimes dx^{\mu_3}\otimes dx^{\mu_4}. \tag{49.25}$$

Taking the exterior derivative of (49.25), we obtain

$$d\mathbf{D}_1\, d\Upsilon = \frac{1}{3!}\delta^{\mu_1\cdots\mu_4}_{\gamma_1\cdots\gamma_4}\frac{\partial}{\partial x^{\mu_1}}\left[(-\Sigma)^{1/2}\Sigma^{\alpha\nu}\frac{\partial\Upsilon}{\partial x^\nu}\right]\varepsilon_{\alpha\mu_2\mu_3\mu_4}dx^{\gamma_1}\otimes\cdots\otimes dx^{\gamma_4} \tag{49.26}$$

Then from (49.23) with $r = 4$ and $\mathbf{\Phi} = d\mathbf{D}_1\, d\Upsilon$,

$$\begin{aligned}
\mathbf{D}_4\, d\mathbf{D}_1\, d\Upsilon &= \frac{1}{4!}\,\frac{1}{3!}(-\Sigma)^{1/2}\Sigma^{\beta_1\gamma_1}\cdots\Sigma^{\beta_4\gamma_4}\varepsilon_{\beta_1\cdots\beta_4}\delta^{\mu_1\cdots\mu_4}_{\gamma_1\cdots\gamma_4}\varepsilon_{\alpha\mu_2\mu_3\mu_4}\\
&\quad\times\frac{\partial}{\partial x^{\mu_1}}\left[(-\Sigma)^{1/2}\Sigma^{\alpha\nu}\frac{\partial\Upsilon}{\partial x^\nu}\right]\\
&= \frac{1}{4!}\,\frac{1}{3!}(-\Sigma)^{1/2}\frac{1}{\Sigma}\varepsilon^{\gamma_1\cdots\gamma_4}\delta^{\mu_1\cdots\mu_4}_{\gamma_1\cdots\gamma_4}\varepsilon_{\alpha\mu_2\mu_3\mu_4}\\
&\quad\times\frac{\partial}{\partial x^{\mu_1}}\left[(-\Sigma)^{1/2}\Sigma^{\alpha\nu}\frac{\partial\Upsilon}{\partial x^\nu}\right]\\
&= -\frac{1}{3!}\,\frac{1}{(-\Sigma)^{1/2}}\varepsilon^{\mu_1\cdots\mu_4}\varepsilon_{\alpha\mu_2\mu_3\mu_4}\frac{\partial}{\partial x^{\mu_1}}\left[(-\Sigma)^{1/2}\Sigma^{\alpha\nu}\frac{\partial\Upsilon}{\partial x^\nu}\right]\\
&= -\frac{1}{(-\Sigma)^{1/2}}\delta_\alpha{}^{\mu_1}\frac{\partial}{\partial x^{\mu_1}}\left[(-\Sigma)^{1/2}\Sigma^{\alpha\nu}\frac{\partial\Upsilon}{\partial x^\nu}\right]\\
&= -\frac{1}{(-\Sigma)^{1/2}}\,\frac{\partial}{\partial x^\alpha}\left[(-\Sigma)^{1/2}\Sigma^{\alpha\nu}\frac{\partial\Upsilon}{\partial x^\nu}\right]\\
&= -\mathrm{Lap}\,\Upsilon,
\end{aligned} \tag{49.27}$$

where we have used (49.10) and the formulas (20.12), (20.13), and (21.6) in IVT-1. Thus (49.13) is proved.

Note. The duality operators $\mathbf{D}_r$, $r = 0, 1, \ldots, 4$, are often denoted collectively by a single symbol $*$, called the *Hodge* $*$ *operator*. We distinguish the duality operators by an extra subscript r because $\mathbf{D}_1$ and $\mathbf{D}_2$ enter explicitly into the ether relations, and they do not have the same properties; e.g., $\mathbf{D}_1^{-1} = \mathbf{D}_3$ but $\mathbf{D}_2^{-1} = -\mathbf{D}_2$. The composition $\mathbf{D}_{4-r}\, d\mathbf{D}_{r+1}$ together with a particular sign is known as the *codifferential* and is denoted usually by the symbol δ. It is important to note, however, that $\mathscr{E}$ is a Minkowskian manifold, not a Riemannian manifold. Hence the sign must be chosen carefully. For a Riemannian manifold the negative sign in (49.13) may be removed.

Now for any given action density $\mathbf{A}$ we determine the function M by (49.8a), and then we solve the function Υ from (49.14). From the function Υ we determine the 1-form $\mathbf{G}$ by (49.5). Then $\mathbf{G}$ is closed, and its dual

$$\mathbf{P} = \mathbf{D}_1\mathbf{G} \tag{49.28}$$

is an action potential, viz.,

$$d\mathbf{P} = d\mathbf{D}_1\mathbf{G} = d\mathbf{D}_1\, d\Upsilon = -\mathbf{D}_0 M = \mathbf{A}. \tag{49.29}$$

Thus we have shown that an action potential $\mathbf{P}$ satisfying the additional condition (49.4) exists.

Note. Relative to a Lorentz system (x^α) the equation (49.14) reduces to

$$\nabla^2\Upsilon - \frac{1}{c^2}\,\frac{\partial^2\Upsilon}{\partial t^2} = -M, \tag{49.30}$$

which has the same form as the equation (32.21) and may be solved by the retarded potential; cf. (32.19). In general a Lorentz system may not exist. Then from (49.10) the equation (49.14) has the form

$$\frac{1}{(-\Sigma)^{1/2}}\,\frac{\partial}{\partial x^\beta}\left[(-\Sigma)^{1/2}\Sigma^{\alpha\beta}\,\frac{\partial\Upsilon}{\partial x^\alpha}\right] = -M, \tag{49.31}$$

which is a linear second-order hyperbolic partial differential equation. We can obtain a solution of (49.31) by solving a Cauchy problem.

We call a closed 1-form $\mathbf{G}$, which is related to an action potential $\mathbf{P}$ by (49.4a) or, equivalently, by (49.28), a *coaction field* associated with the action density $\mathbf{A}$. A potential function Υ for $\mathbf{G}$ is then called a *coaction potential.* The basic relation (49.28) between an action potential and a coaction field is comparable to the Maxwell–Lorentz ether relation (40.26), which pairs a charge–current potential with an electromagnetic field.

The relation between an action density $\mathbf{A}$ and a coaction field $\mathbf{G}$ is not one-to-one in general. Indeed, suppose that Z is a solution of the homogeneous equation

$$\operatorname{Lap} Z = 0. \tag{49.32}$$

Then the closed 1-form

$$\mathbf{Y} = dZ \tag{49.33}$$

may be regarded as the coaction field associated with the zero action density; i.e., the dual of $\mathbf{Y}$ is a closed 3-form

$$\mathbf{W} = \mathbf{D}_1\mathbf{Y} \tag{49.34}$$

such that

$$d\mathbf{W} = \mathbf{0}. \tag{49.35}$$

Hence if $\mathbf{G}$ is a coaction field associated with the action density $\mathbf{A}$, then $\mathbf{G} + \mathbf{W}$ is also.

Conversely, suppose that $\mathbf{G}$ and $\bar{\mathbf{G}}$ are two coaction fields associated with a given action density $\mathbf{A}$. Then the difference

$$Z = \bar{\Upsilon} - \Upsilon, \tag{49.36}$$

where Υ and $\bar{\Upsilon}$ denote the coaction potentials of the coaction fields $\mathbf{G}$ and $\bar{\mathbf{G}}$, respectively, is a solution of the homogeneous equation (49.32). Consequently,

$$\bar{\mathbf{G}} = d\bar{\Upsilon} = d\Upsilon + dZ = \mathbf{G} + \mathbf{W}, \tag{49.37}$$

where $\mathbf{W}$ is defined by (49.34) with Z given by (49.36). Thus the set of coaction fields associated with a given action density may be characterized by the set of solutions of the homogeneous equation (49.32).

In particular, we can force the relation between $\mathbf{A}$ and $\mathbf{G}$ to be one-to-one on a domain $\mathscr{D}$ by imposing a suitable boundary condition on the coaction potential Υ in such a way that the only solutions of the homogeneous equation (49.32) are the constant solutions. Then $\mathbf{W} = \mathbf{0}$ and $\mathbf{G} = \bar{\mathbf{G}}$.

In a series of lectures on electromagnetism and gravitation Toupin[2] formulated a theory in which he used the terms "gravitational field" and "gravitational potential" to describe the 1-form $\mathbf{G}$ and the 0-form Υ, which we call coaction field and coaction potential, respectively. Toupin called the line integral

$$G[\mathscr{C}] = \int_{\mathscr{C}} \mathbf{G} \tag{49.38}$$

on any oriented curve $\mathscr{C}$ the "gravity" of $\mathscr{C}$. From (49.4b) the "gravity" satisfies the following conservation law:

$$\int_{\partial\mathscr{S}} \mathbf{G} = \int_{\mathscr{S}} d\mathbf{G} = 0, \tag{49.39}$$

where $\mathscr{S}$ is any 2-dimensional surface in $\mathscr{E}$.

Since Toupin's terms "gravitational field" and "gravitational potential" are not the same as those in Einstein's theory, which we have followed so far in this chapter, we have changed the terms to coaction field and coaction potential, respectively.

In Toupin's theory the event world is represented by a 4-dimensional differentiable manifold $\mathscr{E}$ together with two basic classes of differential forms which satisfy certain axioms. The first class of differential forms contains the action density $\mathbf{A}$, the action potential $\mathbf{P}$, the "gravitational field" $\mathbf{G}$, and the "gravitational potential" Υ. The axioms governing this class are two conservation laws which correspond to the field equations (49.2) and (49.4b) and an ether relation of the form

$$\mathbf{P} = \mathbf{N}\mathbf{G}, \tag{49.40}$$

where $\mathbf{N} = \mathbf{N}(\mathbf{w})$ is defined at each point $\mathbf{w} \in \mathscr{E}$ as a linear operator from the space of skew-symmetric covariant tensors of order 1 to the space of skew-symmetric tensors of order 3 at the point $\mathbf{w}$. Since a Minkowskian metric and an orientation are not assumed in Toupin's theory, the duality operator $\mathbf{D}_1$ is not defined. Instead, the operator $\mathbf{N}$ is required to possess certain basic properties, which we shall consider in detail in the next section.

It turns out that if a Minkowskian metric and an orientation are defined on $\mathscr{E}$ as in Einstein's theory, then the duality operator $\mathbf{D}_1$ possesses the required basic properties of the operator $\mathbf{N}$ in Toupin's theory. Con-

[2] R. A. Toupin, "Elasticity and electro-magnetics," in *Non-linear Continuum Theories*, pp. 206–342, C.I.M.E. Conference, Bressanone, Italy, 1965. Coordinators: C. Truesdell and G. Grioli.

versely, if $\mathscr{E}$ is a 4-dimensional differentiable manifold equipped with a field $\mathbf{N}$ possessing the basic properties as required in Toupin's theory, then a unique Minkowskian metric and a unique orientation may be defined on $\mathscr{E}$ in such a way that the duality operator $\mathbf{D}_1$ associated with the metric and the orientation coincides with the operator $\mathbf{N}$. Thus the mathematical structures of the models for the event world in Toupin's theory and in Einstein's theory are essentially equivalent to each other. We shall demonstrate this result in the next section.

The second class of differential forms in Toupin's theory contains the charge–current field $\mathbf{\Psi}$, the charge–current potential $\mathbf{\Gamma}$, the electromagnetic field $\mathbf{\Phi}$, and the electromagnetic potential $\mathbf{\Pi}$. Again, these forms are required to satisfy two conservation laws and an ether relation

$$\mathbf{\Gamma} = \mathbf{M\Phi}, \tag{49.41}$$

where $\mathbf{M} = \mathbf{M}(\mathbf{w})$ is defined at each point $\mathbf{w} \in \mathscr{E}$ as a linear operator on the space of skew-symmetric covariant tensors of order 2 at $\mathbf{w}$. In Einstein's theory $\mathbf{M}$ is just the duality operator $\mathbf{D}_2$, which is defined in terms of the Minkowskian metric and the orientation. In Toupin's theory the operator $\mathbf{M}$ is required to possess certain basic properties, which we shall consider in detail in the next chapter. Again, $\mathbf{D}_2$ possesses the required basic properties of $\mathbf{M}$. Conversely, we can use the condition that $\mathbf{M}$ be the same as $\mathbf{D}_2$ to determine a Minkowskian metric and an orientation on $\mathscr{E}$ to within an arbitrary change of gauge. We shall discuss this result in detail in the next chapter.

50. The Nordström–Toupin Ether Relation and the Minkowskian Metric

As mentioned at the end of the preceding section, in Toupin's theory a linear operator

$$\mathbf{N}(\mathbf{w}): \mathscr{E}_{\mathbf{w}}{}^* \to \mathscr{E}_{\mathbf{w}}{}^* \wedge \mathscr{E}_{\mathbf{w}}{}^* \wedge \mathscr{E}_{\mathbf{w}}{}^* \tag{50.1}$$

is defined at each point $\mathbf{w}$ in the 4-dimensional differentiable manifold $\mathscr{E}$, which represents the event world. This operator transforms a coaction field $\mathbf{G}$ into an action potential $\mathbf{P}$,

$$\mathbf{P} = \mathbf{NG}. \tag{50.2}$$

Since the preceding relation is similar to the Maxwell–Lorentz ether relation,

I call it the *Nordström–Toupin ether relation*. In this section we analyze the problem of finding a Minkowskian metric and an orientation on $\mathscr{E}$ such that the condition

$$\mathbf{N} = \mathbf{D}_1, \tag{50.3}$$

is satisfied, where $\mathbf{D}_1$ denotes the duality operator associated with the metric and the orientation.

We prove first the following result:

Uniqueness Theorem (Toupin[3]). If an appropriate Minkowskian metric σ and an orientation on $\mathscr{E}$ may be chosen in such a way that (50.3) holds, then they are unique.

This theorem asserts that the relation between the pair $(\sigma(\mathbf{w}), \mathbf{\Xi}(\mathbf{w}))$ and the duality operator $\mathbf{D}_1(\mathbf{w})$ is one-to-one at each point $\mathbf{w} \in \mathscr{E}$. It is understood that $\mathbf{\Xi}(\mathbf{w})$ denotes the positive unit density tensor associated with $\sigma(\mathbf{w})$ and the particular orientation on $\mathscr{E}_{\mathbf{w}}{}^*$; i.e.,

$$\mathbf{\Xi}(\mathbf{w}) = \mathbf{e}^1(\mathbf{w}) \wedge \mathbf{e}^2(\mathbf{w}) \wedge \mathbf{e}^3(\mathbf{w}) \wedge \mathbf{e}^4(\mathbf{w}) \tag{50.4}$$

for any positive orthonormal basis $\{\mathbf{e}^\alpha(\mathbf{w})\}$ in $\mathscr{E}_{\mathbf{w}}{}^*$.

From now on we shall consider tensors at a particular point $\mathbf{w}$ only. Hence, for simplicity of writing, we shall suppress the notation $\mathbf{w}$ from the argument of the fields $\mathbf{\Xi}$, $\mathbf{D}_1$, σ, $\mathbf{N}$, etc.

We shall prove the uniqueness theorem in several steps. First, we require that the positive unit density tensor $\mathbf{\Xi}$ be held fixed but let the Minkowskian metric σ be replaced by another Minkowskian metric $\bar{\sigma}$. (Notice that $\bar{\sigma}$ is not entirely independent of σ, since the condition that $\mathbf{\Xi}$ is held fixed requires that $\bar{\Sigma} = \Sigma$ relative to any basis in $\mathscr{E}_{\mathbf{w}}{}^*$, where $\bar{\Sigma}$ and Σ denote the determinant of the component matrices of $\bar{\sigma}$ and σ, respectively.) Under the preceding hypothesis we claim that $\bar{\mathbf{D}}_1 = \mathbf{D}_1$ implies that $\bar{\sigma} = \sigma$.

To prove this assertion, we introduce first another class of duality operators. Using the fixed unit density tensor $\mathbf{\Xi}$ and its dual $\mathbf{\Xi}^*$, the unit volume tensor, we define the class of linear operators

$$\mathbf{\Delta}_r\colon \underbrace{\mathscr{E}_{\mathbf{w}}{}^* \wedge \cdots \wedge \mathscr{E}_{\mathbf{w}}{}^*}_{r} \to \underbrace{\mathscr{E}_{\mathbf{w}} \wedge \cdots \wedge \mathscr{E}_{\mathbf{w}}}_{4-r}, \qquad r = 0, 1, 2, 3, 4, \tag{50.5}$$

[3] R. A. Toupin, "Elasticity and electro-magnetics" (see footnote on page 332), Toupin did not state his result in the form presented here, but his result may be reformulated and converted into the theorem here.

by the condition

$$\langle \mathbf{\Delta}_r(\mathbf{X}), \mathbf{Y} \rangle = \langle \mathbf{X} \wedge \mathbf{Y}, \mathbf{\Xi}^* \rangle \tag{50.6}$$

for all $\mathbf{X} \in \mathscr{E}_\mathbf{w}^* \wedge \cdots \wedge \mathscr{E}_\mathbf{w}^*$ of order r and for all $\mathbf{Y} \in \mathscr{E}_\mathbf{w}^* \wedge \cdots \wedge \mathscr{E}_\mathbf{w}^*$ of order $4 - r$. The condition (50.6) is similar to the condition (40.27) for the operators $\mathbf{D}_r$, except that the Minkowskian inner product σ is replaced by the bracket $\langle\,,\,\rangle$. Since the class of operators $\mathbf{\Delta}_r$ depends only on the unit volume tensor $\mathbf{\Xi}^*$, it is not affected by the change of Minkowskian metric from σ to $\bar{\sigma}$ under the assumption that $\mathbf{\Xi}$ is held fixed.

We now give the duality operators $\mathbf{\Delta}_r$ explicitly in terms of an arbitrary positive unit basis $\{\mathbf{e}_\alpha\}$ in $\mathscr{E}_\mathbf{w}$; i.e., $\{\mathbf{e}_\alpha\}$ satisfies the condition

$$\mathbf{\Xi}^* = \mathbf{e}_1 \wedge \mathbf{e}_2 \wedge \mathbf{e}_3 \wedge \mathbf{e}_4. \tag{50.7}$$

This basis may or may not be a positive orthonormal basis relative to σ or relative to $\bar{\sigma}$. Using (50.7) and (50.6), we verify easily that when $r = 0$

$$\mathbf{\Delta}_0(1) = \mathbf{e}_1 \wedge \mathbf{e}_2 \wedge \mathbf{e}_3 \wedge \mathbf{e}_4 = \mathbf{\Xi}^*; \tag{50.8}$$

when $r = 1$

$$\mathbf{\Delta}_1(\mathbf{e}^1) = \mathbf{e}_2 \wedge \mathbf{e}_3 \wedge \mathbf{e}_4, \tag{50.9a}$$

$$\mathbf{\Delta}_1(\mathbf{e}^2) = \mathbf{e}_3 \wedge \mathbf{e}_1 \wedge \mathbf{e}_4, \tag{50.9b}$$

$$\mathbf{\Delta}_1(\mathbf{e}^3) = \mathbf{e}_1 \wedge \mathbf{e}_2 \wedge \mathbf{e}_4, \tag{50.9c}$$

$$\mathbf{\Delta}_1(\mathbf{e}^4) = -\mathbf{e}_1 \wedge \mathbf{e}_2 \wedge \mathbf{e}_3; \tag{50.9d}$$

when $r = 2$

$$\mathbf{\Delta}_2(\mathbf{e}^1 \wedge \mathbf{e}^2) = \mathbf{e}_3 \wedge \mathbf{e}_4, \tag{50.10a}$$

$$\mathbf{\Delta}_2(\mathbf{e}^2 \wedge \mathbf{e}^3) = \mathbf{e}_1 \wedge \mathbf{e}_4, \tag{50.10b}$$

$$\mathbf{\Delta}_2(\mathbf{e}^3 \wedge \mathbf{e}^1) = \mathbf{e}_2 \wedge \mathbf{e}_4, \tag{50.10c}$$

$$\mathbf{\Delta}_2(\mathbf{e}^1 \wedge \mathbf{e}^4) = -\mathbf{e}_3 \wedge \mathbf{e}_2, \tag{50.10d}$$

$$\mathbf{\Delta}_2(\mathbf{e}^2 \wedge \mathbf{e}^4) = -\mathbf{e}_1 \wedge \mathbf{e}_3, \tag{50.10e}$$

$$\mathbf{\Delta}_2(\mathbf{e}^3 \wedge \mathbf{e}^4) = -\mathbf{e}_2 \wedge \mathbf{e}_1; \tag{50.10f}$$

when $r = 3$

$$\mathbf{\Delta}_3(\mathbf{e}^1 \wedge \mathbf{e}^2 \wedge \mathbf{e}^3) = \mathbf{e}_4, \tag{50.11a}$$

$$\mathbf{\Delta}_3(\mathbf{e}^1 \wedge \mathbf{e}^2 \wedge \mathbf{e}^4) = -\mathbf{e}_3, \tag{50.11b}$$

$$\mathbf{\Delta}_3(\mathbf{e}^2 \wedge \mathbf{e}^3 \wedge \mathbf{e}^4) = -\mathbf{e}_1, \tag{50.11c}$$

$$\mathbf{\Delta}_3(\mathbf{e}^3 \wedge \mathbf{e}^1 \wedge \mathbf{e}^4) = -\mathbf{e}_2; \tag{50.11d}$$

and when $r = 4$

$$\mathbf{\Delta}_4(\mathbf{e}^1 \wedge \mathbf{e}^2 \wedge \mathbf{e}^3 \wedge \mathbf{e}^4) = \mathbf{\Delta}_4(\mathbf{\Xi}) = 1. \tag{50.12}$$

Notice that there is a change of sign for terms not containing $\mathbf{e}_4$ on the right-hand side of (50.8)–(50.12) as compared with corresponding terms on the right-hand side of (40.30)–(40.32) and (44.27). This change of sign is due to the fact that in (40.30)–(40.32) and (44.27) $\sigma_1(\mathbf{e}^4, \mathbf{e}^4) = -1$, while in (50.8)–(50.12) $\langle \mathbf{e}_4, \mathbf{e}^4 \rangle = 1$.

Now using the duality operators $\mathbf{\Delta}_1$ and $\mathbf{D}_1$, we verify easily that

$$\sigma_1(\mathbf{X}, \mathbf{Y}) = \langle \mathbf{\Delta}_1(\mathbf{X}), \mathbf{D}_1(\mathbf{Y}) \rangle \tag{50.13}$$

for all $\mathbf{X}$ and $\mathbf{Y}$ in $\mathscr{E}_\mathbf{w}^*$. Indeed, (50.9) and (40.30) are both valid when $\{\mathbf{e}_\alpha\}$ is a positive orthonormal basis relative to σ. Hence if we express $\mathbf{X}$ and $\mathbf{Y}$ in component forms relative to the dual basis $\{\mathbf{e}^\alpha\}$, viz.,

$$\mathbf{X} = X_\alpha \mathbf{e}^\alpha, \qquad \mathbf{Y} = Y_\alpha \mathbf{e}^\alpha, \tag{50.14}$$

then from (50.9) and (40.30)

$$\begin{aligned} \langle \mathbf{\Delta}_1(\mathbf{X}), \mathbf{D}_1(\mathbf{Y}) \rangle &= X_1 Y_1 \langle \mathbf{e}_2 \wedge \mathbf{e}_3 \wedge \mathbf{e}_4, \mathbf{e}^2 \wedge \mathbf{e}^3 \wedge \mathbf{e}^4 \rangle \\ &\quad + X_2 Y_2 \langle \mathbf{e}_3 \wedge \mathbf{e}_1 \wedge \mathbf{e}_4, \mathbf{e}^3 \wedge \mathbf{e}^1 \wedge \mathbf{e}^4 \rangle \\ &\quad + X_3 Y_3 \langle \mathbf{e}_1 \wedge \mathbf{e}_2 \wedge \mathbf{e}_4, \mathbf{e}^1 \wedge \mathbf{e}^2 \wedge \mathbf{e}^4 \rangle \\ &\quad - X_4 Y_4 \langle \mathbf{e}_1 \wedge \mathbf{e}_2 \wedge \mathbf{e}_3, \mathbf{e}^1 \wedge \mathbf{e}^2 \wedge \mathbf{e}^3 \rangle \\ &= X_1 Y_1 + X_2 Y_2 + X_3 Y_3 - X_4 Y_4 = \sigma_1(\mathbf{X}, \mathbf{Y}). \end{aligned} \tag{50.15}$$

Thus (50.13) holds.

Now it follows directly from (50.13) that $\mathbf{D}_1 = \bar{\mathbf{D}}_1$ implies that $\sigma_1 = \bar{\sigma}_1$, or, equivalently, $\sigma = \bar{\sigma}$. Thus the assertion of the first step is proved.

Next, we claim that if only the orientation on $\mathscr{E}_\mathbf{w}$ is held fixed, $\mathbf{D}_1 = \bar{\mathbf{D}}_1$ still implies that $\sigma = \bar{\sigma}$. Notice that when the orientation is held fixed, $\bar{\mathbf{\Xi}}$ is in the same direction of, but need not be equal to, $\mathbf{\Xi}$. Thus the hypothesis of this step is somewhat weaker than that of the previous step.

Since $\mathbf{\Xi}$ and $\bar{\mathbf{\Xi}}$ are in the same direction, they are related by

$$\bar{\mathbf{\Xi}} = a^4 \mathbf{\Xi}, \tag{50.16}$$

where a is a nonzero number. We define another Minkowskian metric $\tilde{\sigma}$ by

$$\tilde{\sigma} = \frac{1}{a^2} \bar{\sigma}. \tag{50.17}$$

Then it can be verified easily that

$$\tilde{\boldsymbol{\Xi}} = \frac{1}{a^4}\,\bar{\boldsymbol{\Xi}} = \boldsymbol{\Xi}, \tag{50.18a}$$

$$\tilde{\mathbf{D}}_1 = \frac{1}{a^2}\,\bar{\mathbf{D}}_1 = \frac{1}{a^2}\,\mathbf{D}_1, \tag{50.18b}$$

where we have used the hypothesis that $\bar{\mathbf{D}}_1 = \mathbf{D}_1$. From (50.18a) we see that $\tilde{\sigma}$ and σ share the same positive unit density tensor. Hence the results of the previous step may be applied.

Specifically, from (50.13)

$$\begin{aligned}\tilde{\sigma}_1(\mathbf{X}, \mathbf{Y}) &= \langle \tilde{\boldsymbol{\Delta}}_1(\mathbf{X}), \tilde{\mathbf{D}}_1(\mathbf{Y})\rangle = \left\langle \boldsymbol{\Delta}_1(\mathbf{X}), \frac{1}{a^2}\,\mathbf{D}_1(\mathbf{Y})\right\rangle \\ &= \frac{1}{a^2}\,\langle \boldsymbol{\Delta}_1(\mathbf{X}), \mathbf{D}_1(\mathbf{Y})\rangle = \frac{1}{a^2}\,\sigma_1(\mathbf{X}, \mathbf{Y})\end{aligned} \tag{50.19}$$

for all $\mathbf{X}$ and $\mathbf{Y}$ in $\mathscr{E}_{\mathbf{w}}{}^*$. Thus

$$\tilde{\sigma}_1 = \frac{1}{a^2}\,\sigma_1, \qquad \tilde{\sigma} = a^2\sigma. \tag{50.20}$$

Combining (50.17) and (50.20), we get

$$\bar{\sigma} = a^4\sigma. \tag{50.21}$$

But then we must have

$$\bar{\mathbf{D}}_1 = a^4\mathbf{D}_1, \tag{50.22}$$

since the orientation on $\mathscr{E}_{\mathbf{w}}$ is held fixed. Clearly (50.22) is consistent with our original hypothesis that $\bar{\mathbf{D}}_1 = \mathbf{D}_1$ only if $a^4 = 1$. Hence we are back to the previous case, and from (50.21) we have $\bar{\sigma} = \sigma$. Thus the assertion of the second step is proved.

Finally, we remove all conditions with regard to the orientation on $\mathscr{E}_{\mathbf{w}}$. We claim that $\bar{\mathbf{D}}_1 = \mathbf{D}_1$ still implies that $\bar{\sigma} = \sigma$. This last step is just the assertion of the uniqueness theorem.

As before we replace σ by $\bar{\sigma}$, but we choose $\bar{\boldsymbol{\Xi}}$ to be in the opposite direction of $\boldsymbol{\Xi}$; i.e.,

$$\bar{\boldsymbol{\Xi}} = -a^4\boldsymbol{\Xi}. \tag{50.23}$$

We show that under the preceding condition the duality operators $\bar{\mathbf{D}}_1$ and $\mathbf{D}_1$ associated with the pairs $(\bar{\sigma}, \bar{\boldsymbol{\Xi}})$ and $(\sigma, \boldsymbol{\Xi})$, respectively, cannot be the same.

Suppose that $\bar{\mathbf{D}}_1$ is the same as $\mathbf{D}_1$. We define $\tilde{\sigma}$ by (50.17), and we choose the orientation of $\tilde{\bar{\Xi}}$ to be the same as that of $\bar{\Xi}$. Then

$$\tilde{\bar{\Xi}} = \frac{1}{a^4}\,\bar{\Xi} = -\Xi, \qquad \tilde{\mathbf{D}}_1 = \frac{1}{a^2}\,\bar{\mathbf{D}}_1 = \frac{1}{a^2}\,\mathbf{D}_1. \tag{50.24}$$

Now since $\tilde{\bar{\Xi}}$ differs from Ξ in sign only, $\{\mathbf{e}_1, \mathbf{e}_2, \mathbf{e}_3, \mathbf{e}_4\}$ is a positive unit basis for Ξ if and only if $\{-\mathbf{e}_1, \mathbf{e}_2, \mathbf{e}_3, \mathbf{e}_4\}$ is a positive unit basis for $\tilde{\bar{\Xi}}$. Then from (50.9)

$$\tilde{\mathbf{\Delta}}_1 = -\mathbf{\Delta}_1. \tag{50.25}$$

Hence from (50.13), (50.24), and (50.25)

$$\begin{aligned}\tilde{\sigma}_1(\mathbf{X}, \mathbf{Y}) &= \langle \tilde{\mathbf{\Delta}}_1(\mathbf{X}), \tilde{\mathbf{D}}_1(\mathbf{Y})\rangle = \left\langle -\mathbf{\Delta}_1(\mathbf{X}), \frac{1}{a^2}\,\mathbf{D}_1(\mathbf{Y})\right\rangle \\ &= \frac{-1}{a^2}\langle \mathbf{\Delta}_1(\mathbf{X}), \mathbf{D}_1(\mathbf{Y})\rangle = \frac{-1}{a^2}\,\sigma_1(\mathbf{X}, \mathbf{Y}).\end{aligned} \tag{50.26}$$

This equation is a contradiction since the negative of a Minkowskian metric is not a Minkowskian metric. Thus the uniqueness theorem is proved. □

The preceding proof of the uniqueness theorem shows clearly that if a Minkowskian metric σ and an orientation on $\mathscr{E}_\mathbf{w}$ may be chosen in such a way that (50.3) holds for a given linear operator $\mathbf{N}$, then that operator cannot be arbitrary but must possess certain basic properties. In order to use these basic properties to determine the pair (σ, Ξ), we must, of course, state them without using the pair (σ, Ξ). For this reason we now choose an arbitrary volume tensor $\bar{\Xi}^*$ in $\mathscr{E}_\mathbf{w} \wedge \mathscr{E}_\mathbf{w} \wedge \mathscr{E}_\mathbf{w} \wedge \mathscr{E}_\mathbf{w}$, and we use that tensor to define the duality operator $\bar{\mathbf{\Delta}}_1$ as before; cf. (50.9). From the operator $\bar{\mathbf{\Delta}}_1$ we define a bilinear form

$$\eta: \mathscr{E}_\mathbf{w}^* \times \mathscr{E}_\mathbf{w}^* \to \mathscr{R} \tag{50.27}$$

by

$$\eta(\mathbf{X}, \mathbf{Y}) = \langle \bar{\mathbf{\Delta}}_1(\mathbf{X}), \mathbf{N}(\mathbf{Y})\rangle \tag{50.28}$$

for all $\mathbf{X}$ and $\mathbf{Y}$ in $\mathscr{E}_\mathbf{w}^*$.

Now we claim that, if there is a pair (σ, Ξ) such that (50.3) holds, the operator $\mathbf{N}$ must possess the following basic property[4]:

(N) Either η or $-\eta$ must be a Minkowskian inner product on $\mathscr{E}_\mathbf{w}^*$.

[4] R. A. Toupin, "Elasticity and electro-magnetics" (see footnote on page 332).

The preceding property is more or less obvious. Indeed, let $\mathbf{N}$ be the same as the duality operator $\mathbf{D}_1$ associated with the pair $(\sigma, \boldsymbol{\Xi})$, and suppose that the density tensor $\bar{\boldsymbol{\Xi}}$ reciprocal to the volume tensor $\bar{\boldsymbol{\Xi}}^*$ is related to the density tensor $\boldsymbol{\Xi}$ by (50.16). Then

$$\bar{\boldsymbol{\Delta}}_1 = \frac{1}{a^4}\boldsymbol{\Delta}_1. \tag{50.29}$$

Hence from (50.13) η is related to σ_1 by

$$\begin{aligned}\eta(\mathbf{X}, \mathbf{Y}) &= \left\langle \frac{1}{a^4}\boldsymbol{\Delta}_1(\mathbf{X}), \mathbf{D}_1(\mathbf{Y})\right\rangle = \frac{1}{a^4}\langle \boldsymbol{\Delta}_1(\mathbf{X}), \mathbf{D}_1(\mathbf{Y})\rangle \\ &= \frac{1}{a^4}\sigma_1(\mathbf{X}, \mathbf{Y}).\end{aligned} \tag{50.30}$$

Thus in this case η is a Minkowskian inner product on $\mathscr{E}_{\mathbf{w}}^*$.

Next, suppose that $\bar{\boldsymbol{\Xi}}$ is related to $\boldsymbol{\Xi}$ by (50.23). Then

$$\bar{\boldsymbol{\Delta}}_1 = -\frac{1}{a^4}\boldsymbol{\Delta}_1. \tag{50.31}$$

Hence from (50.13) η is related to σ_1 by

$$\eta = -\frac{1}{a^4}\sigma_1. \tag{50.32}$$

Thus in this case $-\eta$ is a Minkowskian inner product on $\mathscr{E}_{\mathbf{w}}^*$.

Now since the space of density tensors $\mathscr{E}_{\mathbf{w}}^* \wedge \mathscr{E}_{\mathbf{w}}^* \wedge \mathscr{E}_{\mathbf{w}}^* \wedge \mathscr{E}_{\mathbf{w}}^*$ is 1-dimensional, $\bar{\boldsymbol{\Xi}}$ and $\boldsymbol{\Xi}$ are related either by (50.16) or by (50.23). Thus the proof is complete.

It turns out that the basic property (N) is not only necessary but also sufficient for the condition (50.3). Sufficiency of (N) for (50.3) is asserted by the following result.

Existence Theorem (Toupin[5]). If a linear operator $\mathbf{N}$ of the form (50.1) possesses the basic property (N), then there is a pair $(\sigma, \boldsymbol{\Xi})$ such that (50.3) holds.

We assume that η is a Minkowskian inner product on $\mathscr{E}_{\mathbf{w}}^*$. (Otherwise, we simply reverse the direction of $\bar{\boldsymbol{\Xi}}$ to achieve the hypothesis.) This

[5] R. A. Toupin, "Elasticity and electro-magnetics" (see footnote on page 332). This theorem is reformulated from Toupin's original result.

hypothesis fixes the orientation on $\mathscr{E}_{\mathbf{w}}{}^*$ as that which is determined by $\bar{\boldsymbol{\Xi}}$. Using this orientation and the Minkowskian inner product $\tilde{\sigma}_1 \equiv \eta$, we may determine a positive unit density tensor $\tilde{\bar{\boldsymbol{\Xi}}}$, which generally differs from the reciprocal $\bar{\boldsymbol{\Xi}}$ of $\bar{\boldsymbol{\Xi}}^*$ by a positive factor, say,

$$\tilde{\bar{\boldsymbol{\Xi}}} = a^4 \bar{\boldsymbol{\Xi}}. \tag{50.33}$$

Now we define a Minkowskian inner product σ_1 on $\mathscr{E}_{\mathbf{w}}{}^*$ by

$$\sigma_1{}^* \equiv a^4 \tilde{\sigma}_1 = a^4 \eta, \tag{50.34}$$

and we claim that

$$\sigma_1(\mathbf{X}, \mathbf{Y}) = \langle \boldsymbol{\Delta}_1(\mathbf{X}), \mathbf{N}(\mathbf{Y}) \rangle \tag{50.35}$$

for all $\mathbf{X}$ and $\mathbf{Y}$ in $\mathscr{E}_{\mathbf{w}}{}^*$, where $\boldsymbol{\Delta}_1$ is the duality operator induced by the positive unit volume tensor $\boldsymbol{\Xi}^*$ associated with the dual σ of σ_1 and the chosen orientation on $\mathscr{E}_{\mathbf{w}}{}^*$.

To prove the condition (50.35), we observe first that (50.34) is equivalent to

$$\sigma = \frac{1}{a^4} \tilde{\sigma}. \tag{50.36}$$

Then

$$\boldsymbol{\Xi} = \frac{1}{a^8} \tilde{\bar{\boldsymbol{\Xi}}} = \frac{1}{a^4} \bar{\boldsymbol{\Xi}}, \tag{50.37a}$$

$$\boldsymbol{\Xi}^* = a^4 \bar{\boldsymbol{\Xi}}^*, \tag{50.37b}$$

where (50.33) has been used. It follows from (50.37b) and (50.6) that

$$\boldsymbol{\Delta}_1 = a^4 \bar{\boldsymbol{\Delta}}_1. \tag{50.38}$$

Consequently, from (50.38), (50.28), and (50.34),

$$\begin{aligned} \langle \boldsymbol{\Delta}_1(\mathbf{X}), \mathbf{N}(\mathbf{Y}) \rangle &= \langle a^4 \bar{\boldsymbol{\Delta}}_1(\mathbf{X}), \mathbf{N}(\mathbf{Y}) \rangle = a^4 \langle \bar{\boldsymbol{\Delta}}_1(\mathbf{X}), \mathbf{N}(\mathbf{Y}) \rangle \\ &= a^4 \eta(\mathbf{X}, \mathbf{Y}) = \sigma_1(\mathbf{X}, \mathbf{Y}). \end{aligned} \tag{50.39}$$

Thus (50.35) holds.

Now combining (50.35) and (50.13), we get

$$\langle \boldsymbol{\Delta}_1(\mathbf{X}), \mathbf{N}(\mathbf{Y}) \rangle = \langle \boldsymbol{\Delta}_1(\mathbf{X}), \mathbf{D}_1(\mathbf{Y}) \rangle \tag{50.40}$$

for all **X** and **Y** in $\mathscr{E}_w{}^*$. Since $\boldsymbol{\Delta}_1$ is an isomorphism [cf. (50.9)], and since the bracket $\langle\, ,\, \rangle$ is definite [cf. (31.1) in IVT-1], the condition (50.40) implies that

$$\mathbf{N}(\mathbf{Y}) = \mathbf{D}_1(\mathbf{Y}) \tag{50.41}$$

for all **Y** in $\mathscr{E}_w{}^*$. Thus (50.3) holds, and the proof of the existence theorem is complete. □

Combining the existence theorem and the uniqueness theorem, we see that a metrical structure and an orientation on the differentiable manifold $\mathscr{E}$ may be characterized completely by a smooth field of Nordström–Toupin ether tensors **N** possessing the basic property (N). Conversely, if a Minkowskian metric σ and an orientation are given on $\mathscr{E}$, then the field of duality operators $\mathbf{D}_1$ possesses the basic property (N) and satisfies the Nordström–Toupin ether relation (49.28).

In Toupin's theory two basic classes of differential forms and two fields of ether tensors (**N** and **M**) are regarded as primitive concepts and are defined once and for all on the 4-dimensional differential manifold $\mathscr{E}$, which represents the event world. This situation is similar to the assumption that the Minkowskian metric and the Lorentzian orientation are primitive concepts and are defined once and for all in the Minkowskian manifold $\mathscr{E}$ in Einstein's theory. In this section we have only considered Toupin's first class of differential forms and the Nordström–Toupin ether relation. Our results show that the mathematical structure of Toupin's model may be matched with that of Einstein's model by using the condition (50.3). Then the differential form **A** may be matched with the action density $S\Xi$ considered in Section 48, and the other three differential forms **P**, **G**, and Υ may be matched with the action potential, the coaction field, and the coaction potential, respectively, which are determined by the action density as explained in Section 49.

Note. The set of duality operators $\mathbf{D}_r$, $r = 0, 1, 2, 3, 4$, depends on the orientation but not the Lorentzian orientation on $\mathscr{E}$. Indeed, the conditions (40.27), (40.30)–(40.32), and (44.27) are invariant when we replace the basis $\{\mathbf{e}_\alpha\}$ by the basis $\{-\mathbf{e}_\alpha\}$. (For an even-dimensional space $\{\mathbf{e}_\alpha\}$ and $\{-\mathbf{e}_\alpha\}$ are of the same orientation.) Physically, such a change of basis corresponds to a reversal of time and a central reflection of space on $\mathscr{E}_w$. Since such a transformation does not affect the Nordström–Toupin ether relation, a Lorentzian orientation on $\mathscr{E}$ cannot be determined by the field of Nordström–Toupin ether tensors **N**. In Toupin's theory a positive orientation for timelike vectors is determined by using the entropy principle in thermodynamics. Then a Lorentzian orientation similar to that in Einstein's theory may be defined on each tangent space $\mathscr{E}_{\mathbf{w}}$. We have throughout this work excluded the theory of thermodynamics. Hence we cannot consider that aspect of Toupin's theory here.

8

General Relativistic Theory of Electromagnetism

In the general theory of relativity the event world is represented by the Minkowskian manifold $\mathscr{E}$ whose structure is determined by the distribution of the stress-energy-momentum tensors on $\mathscr{E}$ in accord with Einstein's field equations. This distribution is dominated by its intrinsic part, which is due to the presence of proper mass in a material medium. The electromagnetic field, like the stress tensor field in a material medium, gives rise to only a small contribution in the stress-energy-momentum distribution. Hence we may regard the Minkowskian metric and the electromagnetic field as independent fields on $\mathscr{E}$. In this sense the metrical structure and the orientation on $\mathscr{E}$ are determined to within an arbitrary change of gauge by the Maxwell–Lorentz ether relation. A rigorous theory of electromagnetism in the context of general relativity, however, must allow the electromagnetic field and the gravitational field to affect each other. One exact solution of the coupled system of electromagnetic field equations and gravitational field equations is summarized in the last section of this chapter.

51. Maxwell's Equations in General Relativistic Form

In Chapter 5 we developed the mathematical model for the event world in the special theory of relativity. We recall that the underlying point set of the model is a 4-dimensional affine space $\mathscr{E}$ whose translation space is equipped with a Minkowskian inner product σ and a Lorentzian orientation. We call that model a Minkowskian space–time.

The structure of a Minkowskian space–time may be characterized by a Lorentz frame $\{\mathbf{e}_\alpha\}$, which is an orthonormal basis with respect to σ such that the spacelike basis $\{\mathbf{e}_i\}$ is right handed and the timelike basis vector $\mathbf{e}_4$ points into the future. If we single out a particular point $\mathbf{z}$ in $\mathscr{E}$ as the origin, then a Lorentz frame gives rise to a Lorentz system (x^α), which is an affine coordinate system defined by

$$\mathbf{w} - \mathbf{z} = x^\alpha \mathbf{e}_\alpha \tag{51.1}$$

for all $\mathbf{w}$ in $\mathscr{E}$. A Lorentz frame and a Lorentz system are not unique, of course; transformations among Lorentz frames and Lorentz systems have been discussed in detail in Section 38.

We have pointed out that, when we replace the classical model by the special relativistic model, an inertial frame is replaced by a Lorentz frame and an inertial system is replaced by a Lorentz system. Since the ether frame in the classical theory of electromagnetism is assumed to be an inertial frame, it, too, is replaced by a Lorentz frame. Then we define the electromagnetic field $\mathbf{\Phi}$, the electromagnetic potential $\mathbf{\Pi}$, the charge–current field $\mathbf{\Psi}$, and the charge–current potential $\mathbf{\Gamma}$ by the component formulas (40.8), (40.12), (40.15), and (40.16), respectively, relative to any Lorentz system (x^α) on $\mathscr{E}$. Using these component formulas, we have shown that Maxwell's equations are just the coordinate representations in a Lorentz system for the following field equations:

$$d\mathbf{\Phi} = \mathbf{0}, \tag{51.2a}$$

$$\mathbf{\Phi} = d\mathbf{\Pi}, \tag{51.2b}$$

$$d\mathbf{\Psi} = \mathbf{0}, \tag{51.2c}$$

$$\mathbf{\Psi} = \frac{1}{4\pi}\, d\mathbf{\Gamma}. \tag{51.2d}$$

An important property of these equations is that they are entirely independent of the choice of the Lorentz system. As a result, in special relativity Maxwell's equations are valid in all Lorentz systems. This property, which has been confirmed by experiments, is the very reason that a Minkowskian space–time is chosen as a model for the event world in the special theory of relativity.

Now in the general theory of relativity the model for the event world is a Minkowskian manifold $\mathscr{E}$, which is a 4-dimensional differentiable manifold equipped with a Minkowskian metric σ and a Lorentzian orientation. The tangent space $\mathscr{E}_\mathbf{w}$ of $\mathscr{E}$ at any point $\mathbf{w} \in \mathscr{E}$ has the same structure as the translation space of a Minkowskian space–time. Hence we define a Lorentz frame $\{\mathbf{e}_\alpha(\mathbf{w})\}$ at $\mathbf{w}$ in the same way, namely, $\{\mathbf{e}_\alpha(\mathbf{w})\}$ is an orthonormal basis for $\mathscr{E}_\mathbf{w}$ with respect to $\sigma(\mathbf{w})$ such that $\{\mathbf{e}_i(\mathbf{w})\}$ is right handed and $\mathbf{e}_4(\mathbf{w})$ points into the future. For different points in $\mathscr{E}$ the tangent spaces are independent of one another, however, since $\mathscr{E}$ is not equipped with an affine parallelism. Hence a Lorentz system, which is defined in a Minkowskian space–time by (51.1) on the basis of an affine parallelism, may be defined locally in a Minkowskian manifold only if the Minkowskian metric σ is flat. We have discussed that result in Section 44.

To generalize the theory of electromagnetism from special relativity to general relativity, we replace a Lorentz frame $\{\mathbf{e}_\alpha\}$ for the Minkowskian space–time by a Lorentz frame $\{\mathbf{e}_\alpha(\mathbf{w})\}$ at each point $\mathbf{w}$ in $\mathscr{E}$. More specifically, we define the values $\mathbf{\Phi}(\mathbf{w})$, $\mathbf{\Pi}(\mathbf{w})$, $\mathbf{\Psi}(\mathbf{w})$, and $\mathbf{\Gamma}(\mathbf{w})$ at $\mathbf{w}$ by the component formulas

$$\begin{aligned}\mathbf{\Phi}(\mathbf{w}) &= \big(E^1(\mathbf{w})\mathbf{e}^1(\mathbf{w}) + E^2(\mathbf{w})\mathbf{e}^2(\mathbf{w}) + E^3(\mathbf{w})\mathbf{e}^3(\mathbf{w})\big) \wedge \mathbf{e}^4(\mathbf{w}) \\ &\quad + \big(B^1(\mathbf{w})\mathbf{e}^2(\mathbf{w}) \wedge \mathbf{e}^3(\mathbf{w}) + B^2(\mathbf{w})\mathbf{e}^3(\mathbf{w}) \wedge \mathbf{e}^1(\mathbf{w}) \\ &\quad + B^3(\mathbf{w})\mathbf{e}^1(\mathbf{w}) \wedge \mathbf{e}^2(\mathbf{w})\big), \end{aligned} \tag{51.3}$$

$$\mathbf{\Pi}(\mathbf{w}) = A^1(\mathbf{w})\mathbf{e}^1(\mathbf{w}) + A^2(\mathbf{w})\mathbf{e}^2(\mathbf{w}) + A^3(\mathbf{w})\mathbf{e}^3(\mathbf{w}) + \zeta(\mathbf{w})\mathbf{e}^4(\mathbf{w}), \tag{51.4}$$

$$\begin{aligned}\mathbf{\Psi}(\mathbf{w}) &= \bigg[\frac{j^1(\mathbf{w})}{c}\,\mathbf{e}^2(\mathbf{w}) \wedge \mathbf{e}^3(\mathbf{w}) + \frac{j^2(\mathbf{w})}{c}\,\mathbf{e}^3(\mathbf{w}) \wedge \mathbf{e}^1(\mathbf{w}) \\ &\quad + \frac{j^3(\mathbf{w})}{c}\,\mathbf{e}^1(\mathbf{w}) \wedge \mathbf{e}^2(\mathbf{w})\bigg] \wedge \mathbf{e}^4(\mathbf{w}) \\ &\quad - q(\mathbf{w})\mathbf{e}^1(\mathbf{w}) \wedge \mathbf{e}^2(\mathbf{w}) \wedge \mathbf{e}^3(\mathbf{w}), \end{aligned} \tag{51.5}$$

and

$$\begin{aligned}\mathbf{\Gamma}(\mathbf{w}) &= \big(H^1(\mathbf{w})\mathbf{e}^1(\mathbf{w}) + H^2(\mathbf{w})\mathbf{e}^2(\mathbf{w}) + H^3(\mathbf{w})\mathbf{e}^3(\mathbf{w})\big) \wedge \mathbf{e}^4(\mathbf{w}) \\ &\quad - \big(D^1(\mathbf{w})\mathbf{e}^2(\mathbf{w}) \wedge \mathbf{e}^3(\mathbf{w}) + D^2(\mathbf{w})\mathbf{e}^3(\mathbf{w}) \wedge \mathbf{e}^1(\mathbf{w}) \\ &\quad + D^3(\mathbf{w})\mathbf{e}^1(\mathbf{w}) \wedge \mathbf{e}^2(\mathbf{w})\big), \end{aligned} \tag{51.6}$$

where $\{\mathbf{e}^\alpha(\mathbf{w})\}$ denotes the dual basis of a Lorentz frame $\{\mathbf{e}_\alpha(\mathbf{w})\}$ at $\mathbf{w}$. The component formulas (51.3), (51.4), (51.5), and (51.6) are formally the same as the component formulas (40.8), (40.12), (40.15), and (40.16), respectively, except that the basis $\{\mathbf{e}_\alpha\}$ is replaced by the basis $\{\mathbf{e}_\alpha(\mathbf{w})\}$. We may apply the component formulas (51.3)–(51.6) at each point $\mathbf{w}$ in $\mathscr{E}$. Thus $\mathbf{\Pi}$ is a 1-form, $\mathbf{\Phi}$ and $\mathbf{\Gamma}$ are 2-forms, and $\mathbf{\Psi}$ is a 3-form on $\mathscr{E}$.

The differential forms $\mathbf{\Phi}$, $\mathbf{\Pi}$, $\mathbf{\Psi}$, and $\mathbf{\Gamma}$ are required to satisfy the field equations (51.2) as in special relativity, except that the exterior derivatives are now taken in the Minkowskian manifold $\mathscr{E}$. As we have explained in Section 44, the equation (51.2a) is equivalent to the requirement that

$$\int_{\partial\mathscr{U}} \mathbf{\Phi} = \int_{\mathscr{U}} d\mathbf{\Phi} = 0 \tag{51.7}$$

for all oriented 3-dimensional domains $\mathscr{U}$ in $\mathscr{E}$. The preceding condition corresponds to the conservation of magnetic flux. Next, the equation (51.2c) is equivalent to the requirement that

$$\int_{\partial\mathscr{D}} \mathbf{\Psi} = \int_{\mathscr{D}} d\mathbf{\Psi} = 0 \tag{51.8}$$

for all oriented 4-dimensional domains $\mathscr{D}$ in $\mathscr{E}$. The preceding condition corresponds to the conservation of electric charge. The field equations (51.2b) and (51.2d) are just the potential equations associated with the field equations (51.2a) and (51.2c), respectively, and the potentials $\mathbf{\Pi}$ and $(1/4\pi)\mathbf{\Gamma}$ are unique to within additive closed forms. As remarked in the classical theory and the special theory of relativity, we may determine a particular pair of potentials such that the Maxwell–Lorentz ether relation is satisfied.

Specifically, we use the Lorentz electron theory, which requires that at each point $\mathbf{w}$ in $\mathscr{E}$ the tensors $\mathbf{\Phi}(\mathbf{w})$ and $\mathbf{\Gamma}(\mathbf{w})$ obey the relation

$$\mathbf{\Gamma}(\mathbf{w}) = [\mathbf{D}_2(\mathbf{w})](\mathbf{\Phi}(\mathbf{w})), \tag{51.9}$$

where $\mathbf{D}_2(\mathbf{w})$ denotes the duality operator associated with the Minkowskian inner product and the orientation on $\mathscr{E}_{\mathbf{w}}$. As before we call (51.9) the Maxwell–Lorentz ether relation. Notice that (51.9) is a local condition which is not affected by the curvature of the Minkowskian metric. If we use the component formulas (51.3), (51.6), and (40.31) for $\mathbf{\Phi}(\mathbf{w})$, $\mathbf{\Gamma}(\mathbf{w})$, and $\mathbf{D}_2(\mathbf{w})$, respectively, then we can express (51.9) in component form,

$$D^i(\mathbf{w}) = E^i(\mathbf{w}), \qquad H^i(\mathbf{w}) = B^i(\mathbf{w}), \qquad i = 1, 2, 3, \tag{51.10}$$

which corresponds to the system of ether relations of a vacuum in the classical theory.

Now using an arbitrary positive coordinate system (x^α), we express the differential forms $\mathbf{\Phi}$ and $\mathbf{\Gamma}$ by the component forms

$$\mathbf{\Phi} = \frac{1}{2!}\Phi_{\alpha\beta}\mathbf{h}^\alpha \wedge \mathbf{h}^\beta, \qquad \mathbf{\Gamma} = \frac{1}{2!}\Gamma_{\alpha\beta}\mathbf{h}^\alpha \wedge \mathbf{h}^\beta, \tag{51.11}$$

where $\{\mathbf{h}^\alpha\}$ denotes the natural basis of (x^α) as usual. From (49.23) the ether relation (51.9) may be expressed by the component form

$$\Gamma_{\alpha\beta} = \frac{1}{2!}(-\Sigma)^{1/2}\Sigma^{\mu\nu}\Sigma^{\gamma\theta}\varepsilon_{\mu\gamma\alpha\beta}\Phi_{\nu\theta}. \tag{51.12}$$

The preceding formula may be rewritten in terms of the contravariant components $\Gamma^{\alpha\beta}$ of $\mathbf{\Gamma}$, viz.,

$$\begin{aligned}\Gamma^{\alpha\beta} &= \frac{1}{2!}(-\Sigma)^{1/2}\Sigma^{\mu\nu}\Sigma^{\gamma\theta}\Sigma^{\alpha\lambda}\Sigma^{\beta\omega}\varepsilon_{\mu\gamma\lambda\omega}\Phi_{\nu\theta} \\ &= \frac{1}{2!}(-\Sigma)^{1/2}\frac{1}{\Sigma}\varepsilon^{\nu\theta\alpha\beta}\Phi_{\nu\theta} = -\frac{1}{2!(-\Sigma)^{1/2}}\varepsilon^{\alpha\beta\nu\theta}\Phi_{\nu\theta}\end{aligned} \tag{51.13}$$

where we have used the formula (21.6) in IVT-1. Since the inverse operator of $\mathbf{D}_2$ is just $-\mathbf{D}_2$ [cf. (40.31)], (51.13) is equivalent to

$$\Phi^{\alpha\beta} = \frac{1}{2(-\Sigma)^{1/2}}\varepsilon^{\alpha\beta\nu\theta}\Gamma_{\nu\theta}. \tag{51.14}$$

In the coordinate system (x^α) the system of field equations (51.2) takes the coordinate forms

$$\frac{\partial\Phi_{\alpha\beta}}{\partial x^\gamma} + \frac{\partial\Phi_{\beta\gamma}}{\partial x^\alpha} + \frac{\partial\Phi_{\gamma\alpha}}{\partial x^\beta} = 0, \tag{51.15a}$$

$$\Phi_{\alpha\beta} = \frac{\partial\Pi_\beta}{\partial x^\alpha} - \frac{\partial\Pi_\alpha}{\partial x^\beta}, \tag{51.15b}$$

$$\frac{\partial\Psi_{\alpha\beta\gamma}}{\partial x^\mu} - \frac{\partial\Psi_{\beta\gamma\mu}}{\partial x^\alpha} + \frac{\partial\Psi_{\gamma\mu\alpha}}{\partial x^\beta} - \frac{\partial\Psi_{\mu\alpha\beta}}{\partial x^\gamma} = 0, \tag{51.15c}$$

$$4\pi\Psi_{\alpha\beta\gamma} = \frac{\partial\Gamma_{\beta\gamma}}{\partial x^\alpha} + \frac{\partial\Gamma_{\gamma\alpha}}{\partial x^\beta} + \frac{\partial\Gamma_{\alpha\beta}}{\partial x^\gamma}. \tag{51.15d}$$

These equations may be transformed into more familiar forms in the following way: First, we introduce the dual 1-form $\mathbf{J}$ of $\mathbf{\Psi}$ as before by

$$\mathbf{J} = \mathbf{D}_3\mathbf{\Psi}; \tag{51.16}$$

cf. (41.5). Then from (49.23) the components of $\mathbf{J}$ are given by

$$J_\lambda = \frac{1}{3!}(-\Sigma)^{1/2}\Sigma^{\alpha\mu}\Sigma^{\beta\nu}\Sigma^{\gamma\theta}\Psi_{\alpha\beta\gamma}\varepsilon_{\mu\nu\theta\lambda}, \tag{51.17a}$$

$$J_\lambda = \frac{-1}{3!(-\Sigma)^{1/2}}\Sigma_{\kappa\lambda}\varepsilon^{\alpha\beta\gamma\kappa}\Psi_{\alpha\beta\gamma}, \tag{51.17b}$$

where we have used the formula (21.6) in IVT-1 again. The component formula (51.17b) is equivalent to

$$J^\varkappa = \frac{1}{3!(-\Sigma)^{1/2}} \varepsilon^{\varkappa\alpha\beta\gamma} \Psi_{\alpha\beta\gamma}, \tag{51.18}$$

and the inverse of (51.18) is

$$\Psi_{\alpha\beta\gamma} = (-\Sigma)^{1/2} J^\varkappa \varepsilon_{\varkappa\alpha\beta\gamma}. \tag{51.19}$$

Now in terms of **J** and **Φ** the field equation (51.15c) may be rewritten as

$$\frac{\partial}{\partial x^\varkappa} [(-\Sigma)^{1/2} J^\varkappa] = 0, \tag{51.20}$$

and the field equation (51.15d) may be rewritten as

$$4\pi J^\varkappa = \frac{1}{(-\Sigma)^{1/2}} \frac{\partial}{\partial x^\lambda} [(-\Sigma)^{1/2} \Phi^{\varkappa\lambda}]. \tag{51.21}$$

We can express the preceding equations by using the covariant derivative relative to the Minkowskian metric σ also, viz.,

$$J^\varkappa{}_{,\varkappa} = 0, \qquad \mathrm{Div}\, \mathbf{J} = 0, \tag{51.22}$$

and

$$4\pi J^\varkappa = \Phi^{\varkappa\lambda}{}_{,\lambda} \tag{51.23a}$$

$$4\pi \mathbf{J} = \mathrm{Div}\, \mathbf{\Phi}. \tag{51.23b}$$

Next, using the inverse relation (51.14), we can rewrite the field equation (51.15a) as

$$\frac{\partial}{\partial x^\lambda} [(-\Sigma)^{1/2} \Gamma^{\varkappa\lambda}] = 0, \tag{51.24}$$

which is equivalent to

$$\Gamma^{\varkappa\lambda}{}_{,\lambda} = 0, \qquad \mathrm{Div}\, \mathbf{\Gamma} = \mathbf{0}. \tag{51.25}$$

Finally, the field equation (51.15b) may be rewritten as

$$\Phi_{\alpha\beta} = \Pi_{\beta,\alpha} - \Pi_{\alpha,\beta}, \tag{51.26}$$

where we have used the symmetry of the Christoffel symbols $\left\{ {\alpha \atop \beta\gamma} \right\}$ with respect to (β, γ).

Having rewritten the field equations (51.15a)–(51.15d) in the more familiar forms (51.25), (51.26), (51.22), and (51.23), respectively, we now show that they may be solved in essentially the same way as in the classical theory and in the special theory of relativity. First, we observe as before the fact that the potential $\mathbf{\Pi}$ is unique to within an additive 1-form only. We claim that there is a potential which satisfies the following Lorentz condition:

$$\Pi^{\alpha}{}_{,\alpha} = \operatorname{Div} \mathbf{\Pi} = 0. \tag{51.27}$$

Suppose that $\mathbf{\Pi}$ is any potential which fails to satisfy the preceding condition. Then we may replace $\mathbf{\Pi}$ by

$$\bar{\mathbf{\Pi}} = \mathbf{\Pi} + dZ \tag{51.28}$$

for some function Z, which may be chosen in such a way as to make $\bar{\mathbf{\Pi}}$ satisfy the Lorentz condition (51.27). Indeed, Z may be obtained by solving the generalized Poisson equation

$$\operatorname{Div}(dZ) = \operatorname{Lap}(Z) = -\operatorname{Div} \mathbf{\Pi}, \tag{51.29}$$

where the Laplace operator Lap has been defined by (49.9). Thus without loss of generality we can assume that $\mathbf{\Pi}$ satisfies (51.27).

Now we take the covariant derivative of (51.26) with respect to x^{β} and sum the result as usual, obtaining

$$\Phi^{\alpha\beta}{}_{,\beta} = \Pi^{\beta}{}_{,\alpha\beta} - \Sigma^{\beta\gamma}\Pi^{\alpha}{}_{,\beta\gamma}. \tag{51.30}$$

From the Ricci identities [cf. (59.9) in IVT-2]

$$-\Pi^{\gamma}S_{\gamma\alpha} = \Pi^{\beta}{}_{,\alpha\beta} - \Pi^{\beta}{}_{,\beta\alpha}, \tag{51.31a}$$

$$= \Pi^{\beta}{}_{,\alpha\beta}, \tag{51.31b}$$

where in (51.31b) we have used the Lorentz condition (51.27) to eliminate the term $\Pi^{\beta}{}_{,\beta\alpha}$. Combining (51.30), (51.31b), and (51.23a), we get

$$\Sigma^{\beta\alpha}\Pi_{\alpha,\beta\gamma} + S_{\gamma\alpha}\Pi^{\gamma} + 4\pi J_{\alpha} = 0, \qquad \alpha = 1, 2, 3, 4, \tag{51.32}$$

which is the governing system of equations for the potential $\mathbf{\Pi}$.

As we have remarked in Section 45, the electromagnetic field contributes to only a small extra part of the stress-energy-momentum tensor $\mathbf{\Theta}$ on $\mathscr{E}$. Hence in the system (51.32) we may regard the Ricci tensor $\mathbf{S}$ as a given field on $\mathscr{E}$ independent of Π^{γ} and J_{α}. Then we can solve the system (51.32) for Π^{γ} subject to the Lorentz condition (51.27) for each given 1-form $\mathbf{J}$.

(Recall that $\mathbf{J}$ is just the dual of the 3-form $\boldsymbol{\Psi}$.) After obtaining the potential $\boldsymbol{\Pi}$, we can determine $\boldsymbol{\Phi}$ by (51.26) and then $\boldsymbol{\Gamma}$ by (51.12).

The preceding analysis of the field equations (51.15) is similar to that of the field equations (49.2), (49.3), (49.4b), and (49.5), which govern the forms $\mathbf{A}$, $\mathbf{P}$, $\mathbf{G}$, and Υ. Here we regard the charge–current field $\boldsymbol{\Psi}$ as a given closed 3-form on $\mathscr{E}$, and we determine the corresponding forms $\boldsymbol{\Phi}$, $\boldsymbol{\Pi}$, and $\boldsymbol{\Gamma}$ in such a way that the ether relation (51.9) and the field equations (51.15) are satisfied. In Section 49 the action density $\mathbf{A}$ is a given 4-form on $\mathscr{E}$, and we determine the corresponding forms $\mathbf{G}$, Υ, and $\mathbf{P}$ in such a way that the ether relation (49.4a) and the field equations (49.2)–(49.5) are satisfied.

It should be noted that both the analysis in Section 49 and the analysis here are based on a given Minkowskian metric and a given orientation on $\mathscr{E}$. Indeed, the duality operators $\mathbf{D}_1$ and $\mathbf{D}_2$, which characterize the ether relations, are induced by the metric and the orientation. In general relativity the metric cannot really be chosen *a priori*, however, since it must satisfy Einstein's field equations, which depend on the stress-energy-momentum tensor $\boldsymbol{\Theta}$ on $\mathscr{E}$. Because a small extra part of $\boldsymbol{\Theta}$ depends on the electromagnetic field and the extra part of the action density, the field equations of electromagnetism and action may be analyzed rigorously only if they are coupled with Einstein's field equations. An example of a rigorous solution of the coupled system of Einstein's field equations and Maxwell's field equations is summarized in the last section of this chapter.

52. The Maxwell–Lorentz Ether Relation and the Minkowskian Metric I: Toupin's Uniqueness Theorem

In Section 50 we have shown that the structure induced on a 4-dimensional differentiable manifold $\mathscr{E}$ by a field of Nordström–Toupin ether tensors $\mathbf{N}$, which possess a certain basic property (N), is essentially equivalent to the structure induced by a Minkowskian metric and an orientation on the manifold. In the next three sections we shall prove a similar result for the field of Maxwell–Lorentz ether tensors.

We recall first that a Maxwell–Lorentz ether tensor $\mathbf{M}(\mathbf{w})$ at a point $\mathbf{w}$ in $\mathscr{E}$ is a linear operator

$$\mathbf{M}(\mathbf{w})\colon \mathscr{E}_{\mathbf{w}}^* \wedge \mathscr{E}_{\mathbf{w}}^* \to \mathscr{E}_{\mathbf{w}}^* \wedge \mathscr{E}_{\mathbf{w}}^* \tag{52.1}$$

such that

$$\boldsymbol{\Gamma}(\mathbf{w}) = [\mathbf{M}(\mathbf{w})](\boldsymbol{\Phi}(\mathbf{w})), \tag{52.2}$$

where $\mathbf{\Gamma}(\mathbf{w})$ and $\mathbf{\Phi}(\mathbf{w})$ are the values of the charge–current potential and the electromagnetic field at $\mathbf{w}$, as defined in the preceding section. In the general relativistic formulation of the theory of electromagnetism $\mathbf{\Gamma}(\mathbf{w})$ and $\mathbf{\Phi}(\mathbf{w})$ are related by (51.9), which implies that the Maxwell–Lorentz ether tensor at $\mathbf{w}$ is just the duality operator $\mathbf{D}_2(\mathbf{w})$ associated with the Minkowskian inner product $\sigma(\mathbf{w})$ and the orientation on $\mathscr{E}_{\mathbf{w}}$. That condition is a local result, which is not affected by the curvature of the field σ.

In Section 50 we have shown that there is a one-to-one correspondence between the duality operator $\mathbf{D}_1(\mathbf{w})$ and the pair $(\sigma(\mathbf{w}), \Xi(\mathbf{w}))$, which gives rise to $\mathbf{D}_1(\mathbf{w})$; cf. the uniqueness theorem in that section. Now we wish to establish a similar (but somewhat weaker) result for the duality operator $\mathbf{D}_2(\mathbf{w})$.

We observe first that a direct one-to-one correspondence between $\mathbf{D}_2(\mathbf{w})$ and $(\sigma(\mathbf{w}), \Xi(\mathbf{w}))$ is not valid. Indeed, an inspection of the component formula (40.31) for $\mathbf{D}_2$ shows clearly that $\mathbf{D}_2(\mathbf{w})$ is invariant when we replace the positive orthonormal basis $\{\mathbf{e}_\alpha(\mathbf{w})\}$ by a positive basis $\{a\mathbf{e}_\alpha(\mathbf{w})\}$ for any nonzero number a. This simple remark implies that the duality operator $\mathbf{D}_2$ associated with the pair (σ, Ξ) coincides with that associated with the pair $(a^2\sigma, a^4\Xi)$. Consequently, a uniqueness theorem for $\mathbf{D}_2$, worded exactly as the uniqueness theorem in Section 50 for $\mathbf{D}_1$, is not true. However, the following weaker result may be proved.

Uniqueness Theorem (Toupin[(1)]). Let $\mathbf{D}_2(\mathbf{w})$ and $\bar{\mathbf{D}}_2(\mathbf{w})$ denote the duality operators associated with the pairs $(\sigma(\mathbf{w}), \Xi(\mathbf{w}))$ and $(\bar{\sigma}(\mathbf{w}), \bar{\Xi}(\mathbf{w}))$, respectively. Then $\mathbf{D}_2(\mathbf{w})$ coincides with $\bar{\mathbf{D}}_2(\mathbf{w})$ if and only if there is a nonzero number $a(\mathbf{w})$ such that

$$(\bar{\sigma}(\mathbf{w}), \bar{\Xi}(\mathbf{w})) = (a^2(\mathbf{w})\sigma(\mathbf{w}), a^4(\mathbf{w})\Xi(\mathbf{w})). \tag{52.3}$$

Note. Since the preceding theorem is strictly a result for each point $\mathbf{w}$ in $\mathscr{E}$, the number $a(\mathbf{w})$ in (52.3) generally may depend on $\mathbf{w}$. We call a transformation from (σ, Ξ) to $(a^2\sigma, a^4\Xi)$ a change of *gauge*; such a transformation is said to be *uniform* if a is a constant, and *nonuniform* if a depends on $\mathbf{w}$. Then the preceding theorem asserts simply that the duality operator $\mathbf{D}_2$ is invariant under an arbitrary, possibly nonuniform, change of gauge.

(1) R. A. Toupin, Elasticity and electro-magnetics, in *Non-Linear Continuum Theories*, pp. 206–342, C.I.M.E. Conference, Bressanone, Italy, 1965. Coordinators: C. Truesdell and G. Grioli. Toupin did not state his result in the form presented here, but his result may be reformulated and converted into the theorem here.

From now on we shall consider tensors at a particular point $\mathbf{w}$ only. Hence, for simplicity of writing, we shall suppress the notation $\mathbf{w}$ from the argument of the field $\mathbf{M}$, $\mathbf{D}_2$, σ, $\boldsymbol{\Xi}$, etc.

To prove the uniqueness theorem, we establish first the following useful identity:

$$\langle \mathbf{D}_2(\mathbf{f}^1 \wedge \mathbf{f}^2), \boldsymbol{\Delta}_2(\mathbf{g}^1 \wedge \mathbf{g}^2)\rangle = \sigma_1(\mathbf{f}^1, \mathbf{g}^1)\sigma_1(\mathbf{f}^2, \mathbf{g}^2) - \sigma_1(\mathbf{f}^1, \mathbf{g}^2)\sigma_1(\mathbf{f}^2, \mathbf{g}^1), \tag{52.4}$$

which is valid for all $\mathbf{f}^1, \mathbf{f}^2, \mathbf{g}^1, \mathbf{g}^2$ in $\mathscr{E}_{\mathbf{w}}^*$. Here $\boldsymbol{\Delta}_2$ denotes the duality operator associated with the unit density tensor $\boldsymbol{\Xi}$ as explained in Section 50; cf. (50.6) and (50.10). We verify the identity (52.4) by direct calculation based on the component formulas (40.31) for $\mathbf{D}_2$ and (50.10) for $\boldsymbol{\Delta}_2$ relative to a particular positive orthonormal basis $\{\mathbf{e}_\alpha\}$ for $\mathscr{E}_{\mathbf{w}}$.

Specifically, the left-hand side of (52.4) may be expressed as

$$\begin{aligned}\langle \mathbf{D}_2(\mathbf{f}^1 \wedge \mathbf{f}^2), \boldsymbol{\Delta}_2(\mathbf{g}^1 \wedge \mathbf{g}^2)\rangle &= (f_1{}^1 f_2{}^2 - f_1{}^2 f_2{}^1)(g_1{}^1 g_2{}^2 - g_1{}^2 g_2{}^1) \\ &\quad + (f_1{}^1 f_3{}^2 - f_1{}^2 f_3{}^1)(g_1{}^1 g_3{}^2 - g_1{}^2 g_3{}^1) \\ &\quad + (f_2{}^1 f_3{}^2 - f_2{}^2 f_3{}^1)(g_2{}^1 g_3{}^2 - g_2{}^2 g_3{}^1) \\ &\quad - (f_1{}^1 f_4{}^2 - f_1{}^2 f_4{}^1)(g_1{}^1 g_4{}^2 - g_1{}^2 g_4{}^1) \\ &\quad - (f_2{}^1 f_4{}^2 - f_2{}^2 f_4{}^1)(g_2{}^1 g_4{}^2 - g_2{}^2 g_4{}^1) \\ &\quad - (f_3{}^1 f_4{}^2 - f_3{}^2 f_4{}^1)(g_3{}^1 g_4{}^2 - g_3{}^2 g_4{}^1),\end{aligned} \tag{52.5}$$

while the right-hand side may be expressed as

$$\begin{aligned}&\sigma_1(\mathbf{f}^1, \mathbf{g}^1)\sigma_1(\mathbf{f}^2, \mathbf{g}^2) - \sigma_1(\mathbf{f}^1, \mathbf{g}^2)\sigma_1(\mathbf{f}^2, \mathbf{g}^1) \\ &\quad = (f_1{}^1 g_1{}^1 + f_2{}^1 g_2{}^1 + f_3{}^1 g_3{}^1 - f_4{}^1 g_4{}^1)(f_1{}^2 g_1{}^2 + f_2{}^2 g_2{}^2 + f_3{}^2 g_3{}^2 - f_4{}^2 g_4{}^2) \\ &\qquad - (f_1{}^1 g_1{}^2 + f_2{}^1 g_2{}^2 + f_3{}^1 g_3{}^2 - f_4{}^1 g_4{}^2)(f_1{}^2 g_1{}^1 + f_2{}^2 g_2{}^1 + f_3{}^2 g_3{}^1 - f_4{}^2 g_4{}^1).\end{aligned} \tag{52.6}$$

We verify that the right-hand sides of (52.5) and (52.6) are the same by multiplying the factors and rearranging the terms. Thus the identity (52.4) is proved.

Now we follow the same procedure as in the proof of the uniqueness theorem for $\mathbf{D}_1$. First, we hold the unit density tensor $\boldsymbol{\Xi}$ fixed but we change the Minkowskian metric from σ to $\bar{\sigma}$. Under this hypothesis we claim that $\bar{\mathbf{D}}_2 = \mathbf{D}_2$ implies that $\bar{\sigma} = \sigma$.

Indeed, when $\boldsymbol{\Xi}$ is held fixed, the duality operator $\boldsymbol{\Delta}_2$ is also. Hence if $\bar{\mathbf{D}}_2 = \mathbf{D}_2$, then the right-hand side of (52.4) is fixed. Thus

$$\begin{aligned}&\bar{\sigma}_1(\mathbf{f}^1, \mathbf{g}^1)\bar{\sigma}_1(\mathbf{f}^2, \mathbf{g}^2) - \bar{\sigma}_1(\mathbf{f}^1, \mathbf{g}^2)\bar{\sigma}_1(\mathbf{f}^2, \mathbf{g}^1) \\ &\quad = \sigma_1(\mathbf{f}^1, \mathbf{g}^1)\sigma_1(\mathbf{f}^2, \mathbf{g}^2) - \sigma_1(\mathbf{f}^1, \mathbf{g}^2)\sigma_1(\mathbf{f}^2, \mathbf{g}^1).\end{aligned} \tag{52.7}$$

Applying the preceding formula to a positive orthonormal basis $\{\mathbf{e}^\alpha\}$ relative to σ_1, we get

$$\bar{\Sigma}^{\alpha\beta}\bar{\Sigma}^{\gamma\theta} - \bar{\Sigma}^{\alpha\gamma}\bar{\Sigma}^{\beta\theta} = \Delta^{\alpha\beta}\Delta^{\gamma\theta} - \Delta^{\alpha\gamma}\Delta^{\beta\theta}, \tag{52.8}$$

where $[\Delta^{\alpha\beta}]$ denotes the constant matrix diag(1, 1, 1, −1) as before. We wish to show that

$$[\bar{\Sigma}^{\alpha\beta}] = [\Delta^{\alpha\beta}], \tag{52.9}$$

which is just the component form of $\bar{\sigma}_1 = \sigma_1$.

We restrict our attention first to the indices ranging from 1 to 3 in (52.8), viz.,

$$\bar{\Sigma}^{ij}\bar{\Sigma}^{kl} - \bar{\Sigma}^{ik}\bar{\Sigma}^{jl} = \delta^{ij}\delta^{kl} - \delta^{ik}\delta^{jl}. \tag{52.10}$$

Multiplying (52.10) by $\varepsilon_{ikm}\bar{\Sigma}^{mn}$ and summing the result on repeated indices from 1 to 3, we get

$$\varepsilon^{jlp}\delta^{pn}\det[\bar{\Sigma}^{rs}] = \varepsilon^{jlp}\bar{\Sigma}^{pn}, \tag{52.11}$$

or, equivalently,

$$\delta^{pn}\det[\bar{\Sigma}^{rs}] = \bar{\Sigma}^{pn}, \tag{52.12}$$

where we have used the formula (21.6) in IVT-1. Taking the determinant of this matrix equation, we obtain

$$(\det[\bar{\Sigma}^{rs}])^2 = 1. \tag{52.13}$$

Thus (52.12) reduces to

$$\bar{\Sigma}^{ij} = \pm\delta^{ij}. \tag{52.14}$$

We shall show that the sign in (52.14) must be positive.

Next, we restrict our attention to the case that one of the indices in (52.8) is 4, while the remaining three indices range from 1 to 3, viz.,

$$\bar{\Sigma}^{ij}\bar{\Sigma}^{k4} - \bar{\Sigma}^{ik}\bar{\Sigma}^{j4} = 0. \tag{52.15}$$

Substituting the result (52.14) into the preceding equation, we obtain immediately

$$\bar{\Sigma}^{k4} = 0. \tag{52.16}$$

Finally, we choose two of the indices in (52.8) to be 4, viz.,

$$\bar{\Sigma}^{ij}\bar{\Sigma}^{44} - \bar{\Sigma}^{i4}\bar{\Sigma}^{j4} = -\delta^{ij}, \tag{52.17}$$

which implies that

$$\bar{\Sigma}^{44} = -1 \qquad \text{when} \quad \bar{\Sigma}^{ij} = \delta^{ij}, \tag{52.18}$$

and that

$$\bar{\Sigma}^{44} = +1 \qquad \text{when} \quad \bar{\Sigma}^{ij} = -\delta^{ij}. \tag{52.19}$$

Since $\bar{\sigma}$ is required to be a Minkowskian metric, we must rule out (52.19). Thus the sign in (52.14) must be positive, and (52.9) is proved.

Next, we replace the hypothesis that $\mathbf{\Xi}$ be held fixed by the weaker hypothesis that the orientation on $\mathscr{E}_{\mathbf{w}}$ be held fixed. Under this weaker hypothesis we claim that $\bar{\mathbf{D}}_2 = \mathbf{D}_2$ implies that $(\sigma, \mathbf{\Xi})$ and $(\bar{\sigma}, \bar{\mathbf{\Xi}})$ are related by (52.3).

This result may be proved in the same way as in the proof of the uniqueness theorem for $\mathbf{D}_1$. Indeed, since $\mathbf{\Xi}$ and $\bar{\mathbf{\Xi}}$ are in the same direction, they are related by

$$\bar{\mathbf{\Xi}} = a^4 \mathbf{\Xi}; \tag{52.20}$$

cf. (50.16). Then we define the Minkowskian metric $\tilde{\sigma}$ by

$$\tilde{\sigma} = \frac{1}{a^2}\,\bar{\sigma}; \tag{52.21}$$

cf. (50.17). As before (52.21) implies that

$$\tilde{\mathbf{\Xi}} = \frac{1}{a^4}\,\bar{\mathbf{\Xi}} = \mathbf{\Xi} \tag{52.22}$$

cf. (50.18a). Now since the duality operator $\mathbf{D}_2$ is gauge invariant, it follows from (52.21) and (52.22) that $\tilde{\mathbf{D}}_2$ coincides with $\bar{\mathbf{D}}_2$. Hence $\tilde{\mathbf{D}}_2 = \mathbf{D}_2$ and $\tilde{\mathbf{\Xi}} = \mathbf{\Xi}$. Thus the conditions of the previous step are satisfied, and, as a result,

$$\tilde{\sigma} = \sigma. \tag{52.23}$$

Substituting (52.23) and (52.21) into (52.22), we obtain (52.3).

Finally, we remove all conditions with regard to the orientation on $\mathscr{E}_{\mathbf{w}}$. We claim that $\bar{\mathbf{D}}_2 = \mathbf{D}_2$ still implies that $(\sigma, \mathbf{\Xi})$ and $(\bar{\sigma}, \bar{\mathbf{\Xi}})$ are related by (52.3). This last step is just the assertion of the uniqueness theorem.

The proof is again the same as that in the proof of the uniqueness theorem for $\mathbf{D}_1$. We show that, when $\bar{\mathbf{\Xi}}$ is in the opposite direction of $\mathbf{\Xi}$, $\bar{\mathbf{D}}_2$ cannot be the same as $\mathbf{D}_2$. Suppose that $\bar{\mathbf{\Xi}}$ and $\mathbf{\Xi}$ are related by

$$\bar{\mathbf{\Xi}} = -a^4 \mathbf{\Xi}; \tag{52.24}$$

cf. (50.23). We define $\tilde{\sigma}$ as before by (52.21), and we choose the orientation of $\tilde{\bar{\boldsymbol{\Xi}}}$ to be the same as that of $\bar{\boldsymbol{\Xi}}$. Then

$$\tilde{\bar{\boldsymbol{\Xi}}} = \frac{1}{a^4}\,\bar{\boldsymbol{\Xi}} = -\boldsymbol{\Xi}; \tag{52.25}$$

cf. (50.24). Hence $\tilde{\mathbf{D}}_2 = \bar{\mathbf{D}}_2$ as before. Now by hypothesis $\bar{\mathbf{D}}_2 = \mathbf{D}_2$. Thus the duality operators $\tilde{\mathbf{D}}_2$ and $\mathbf{D}_2$ associated with the pairs $(\tilde{\sigma}, -\boldsymbol{\Xi})$ and $(\sigma, \boldsymbol{\Xi})$ coincide. We claim that this condition gives raise to a contradiction.

Indeed, from (50.6) when $\boldsymbol{\Xi}$ is replaced by $-\boldsymbol{\Xi}$, $\boldsymbol{\Delta}_2$ is transformed into $-\boldsymbol{\Delta}_2$. Then the identity (52.4) implies that

$$\begin{aligned} -\tilde{\sigma}_1(\mathbf{f}^1, \mathbf{g}^1)\tilde{\sigma}_1(\mathbf{f}^2, \mathbf{g}^2) &+ \tilde{\sigma}_1(\mathbf{f}^1, \mathbf{g}^2)\tilde{\sigma}_1(\mathbf{f}^2, \mathbf{g}^1) \\ &= \sigma_1(\mathbf{f}^1, \mathbf{g}^1)\sigma_1(\mathbf{f}^2, \mathbf{g}^2) - \sigma_1(\mathbf{f}^1, \mathbf{g}^2)\sigma_1(\mathbf{f}^2, \mathbf{g}^1). \end{aligned} \tag{52.26}$$

Following the procedure from (52.7) to (52.12), we obtain from (52.26)

$$-\delta^{pn}\det[\tilde{\Sigma}^{rs}] = \tilde{\Sigma}^{pn}. \tag{52.27}$$

Now taking the determinant of the preceding matrix equation, we get

$$(\det[\tilde{\Sigma}^{rs}])^2 = -1, \tag{52.28}$$

which is a contradiction. Thus the uniqueness theorem is proved. □

Note. From the preceding uniqueness theorem we see that the Maxwell–Lorentz ether tensor $\mathbf{M}$ must possess certain basic properties; otherwise, it cannot be the duality operator $\mathbf{D}_2$ associated with any pair $(\sigma, \boldsymbol{\Xi})$. We shall consider these basic properties in detail in the following section.

Like the Nordström–Toupin ether relation, the Maxwell–Lorentz ether relation is invariant when we retain the orientation but reverse the Lorentzian orientation on the tangent space $\mathscr{E}_{\mathbf{w}}$. Specifically, the system of formulas (40.31) remains unchanged when we replace the basis $\{\mathbf{e}_\alpha\}$ by the basis $\{-\mathbf{e}_\alpha\}$. Hence a Lorentzian orientation on $\mathscr{E}$ cannot be determined by the tensor field $\mathbf{M}$.

53. The Maxwell–Lorentz Ether Relation and the Minkowskian Metric II: Basic Properties and Preliminary Lemmas

In the preceding section we showed that the duality operator $\mathbf{D}_2$ may determine the orientation and the Minkowskian metric on $\mathscr{E}$ only to within an arbitrary change of gauge. In general if a field of linear maps

$\mathbf{M}$ of the form (52.1) is given on $\mathscr{E}$, there may or may not be any pair (σ, Ξ) such that $\mathbf{M} = \mathbf{D}_2$. In this section we obtain some necessary conditions on $\mathbf{M}$ in order that $\mathbf{M}$ be the same as the duality operator $\mathbf{D}_2$ associated with some pair (σ, Ξ). Then in the following section we show that these conditions on $\mathbf{M}$ are also sufficient for the existence of a pair (σ, Ξ) such that the condition (52.3) holds. A similar problem for the Nordström–Toupin ether tensor $\mathbf{N} = \mathbf{D}_1$ has been solved in Section 50.

Since we shall consider the Maxwell–Lorentz ether tensor $\mathbf{M}(\mathbf{w})$ at a particular point $\mathbf{w}$ in $\mathscr{E}$ only, for simplicity of writing we shall drop from the notation the argument $\mathbf{w}$ as before.

The basic properties[(2)] of the tensor $\mathbf{M}$ are the following:

(M-1) *Reciprocity.* The ether tensor $\mathbf{M}$ obeys the condition

$$\mathbf{M}^2 = -\mathbf{I}_2 \tag{53.1}$$

or, equivalently,

$$\mathbf{M}^{-1} = -\mathbf{M}, \tag{53.2}$$

where $\mathbf{I}_2$ denotes the identity map on $\mathscr{E}_{\mathbf{w}}^* \wedge \mathscr{E}_{\mathbf{w}}^*$.

The proof of this property follows directly from (40.31); i.e.,

$$\mathbf{D}_2{}^{-1} = -\mathbf{D}_2. \tag{53.3}$$

(M-2) *Symmetry.* The bilinear form

$$\mu: \mathscr{E}_{\mathbf{w}}^* \wedge \mathscr{E}_{\mathbf{w}}^* \times \mathscr{E}_{\mathbf{w}}^* \wedge \mathscr{E}_{\mathbf{w}}^* \to \mathscr{R} \tag{53.4}$$

defined by

$$\mu(\mathbf{X}, \mathbf{Y}) \equiv \langle \bar{\mathbf{\Delta}}_2(\mathbf{X}), \mathbf{M}(\mathbf{Y}) \rangle \tag{53.5}$$

is symmetric; i.e.,

$$\mu(\mathbf{X}, \mathbf{Y}) = \mu(\mathbf{Y}, \mathbf{X}) \tag{53.6}$$

for all $\mathbf{X}$ and $\mathbf{Y}$ in $\mathscr{E}_{\mathbf{w}}^* \wedge \mathscr{E}_{\mathbf{w}}^*$. In (53.5) $\bar{\mathbf{\Delta}}_2$ denotes the duality operator induced by a certain nonzero volume tensor $\bar{\Xi}^*$, which is chosen as a reference in order to state the basic property without using any metric and orientation.

The proof of the preceding property follows directly from (50.10) and (40.31). Indeed, let $\mathbf{M}$ be the duality operator $\mathbf{D}_2$ associated with the pair (σ, Ξ). Then Ξ and $\bar{\Xi}$ may or may not be in the same direction. From (40.31) and (50.10) when Ξ and $\bar{\Xi}$ are in the same direction, the bilinear

(2) R. A. Toupin, "Elasticity and electro-magnetics" (see footnote on page 351).

form may be expressed by

$$\mu(\mathbf{X}, \mathbf{Y}) = \frac{1}{a^4}(X_{12}Y_{12} + X_{23}Y_{23} + X_{31}Y_{31} - X_{14}Y_{14} - X_{24}Y_{24} - X_{34}Y_{34}), \tag{53.7}$$

but when $\mathbf{\Xi}$ and $\bar{\mathbf{\Xi}}$ are in the opposite directions, μ is given by

$$\mu(\mathbf{X}, \mathbf{Y}) = \frac{1}{a^4}(-X_{12}Y_{12} - X_{23}Y_{23} - X_{31}Y_{31} + X_{14}Y_{14} + X_{24}Y_{24} + X_{34}Y_{34}), \tag{53.8}$$

where $X_{\alpha\beta}$ and $Y_{\alpha\beta}$ denote the components of $\mathbf{X}$ and $\mathbf{Y}$ relative to a positive orthonormal basis $\{\mathbf{e}_\alpha\}$ with respect to the pair $(\sigma, \mathbf{\Xi})$, and where

$$\bar{\mathbf{\Xi}} = \pm a^4 \mathbf{\Xi}. \tag{53.9}$$

The sign in (53.9) is chosen according to the hypothesis on the directions of $\mathbf{\Xi}$ and $\bar{\mathbf{\Xi}}$. From (53.8) and (53.9) we see clearly that the bilinear form μ is symmetric in either case.

(M-3) *Definiteness*. There is a covariant vector $\mathbf{e}^4$ in $\mathscr{E}_{\mathbf{w}}^*$ such that the quadratic form $\mu(\mathbf{X}, \mathbf{X})$, restricted to all simple tensors in $\mathscr{E}_{\mathbf{w}}^* \wedge \mathscr{E}_{\mathbf{w}}^*$ having $\mathbf{e}^4$ as a factor, is either positive definitive or negative definite, where μ is the bilinear form defined in the basic property (M-2).

The proof of the preceding property follows directly form the component formulas (53.7) and (53.8) for the bilinear form μ in general. Indeed, we choose the particular covariant vector $\mathbf{e}^4$ to be just the fourth basis vector in the positive orthonormal basis $\{\mathbf{e}^\alpha\}$. Then for any nonzero simple tensor $\mathbf{X}$ in $\mathscr{E}_{\mathbf{w}}^* \wedge \mathscr{E}_{\mathbf{w}}^*$ having $\mathbf{e}^4$ as a factor $\mu(\mathbf{X}, \mathbf{X})$ is negative definite, viz.,

$$\mu(\mathbf{X}, \mathbf{X}) = \frac{1}{a^4}(-X_{14}^2 - X_{24}^2 - X_{34}^2) < 0, \tag{53.10}$$

when $\mathbf{\Xi}$ and $\bar{\mathbf{\Xi}}$ are in the same direction, but it is positive definite, viz.,

$$\mu(\mathbf{X}, \mathbf{X}) = \frac{1}{a^4}(X_{14}^2 + X_{24}^2 + X_{34}^2) > 0, \tag{53.11}$$

when $\mathbf{\Xi}$ and $\bar{\mathbf{\Xi}}$ are in the opposite directions. Thus the basic property is necessary for $\mathbf{M}$.

By using the basic property (M-3), we can determine the orientation on $\mathscr{E}_{\mathbf{w}}$ uniquely as follows: We choose $\bar{\mathbf{\Xi}}$ in such a way that the quadratic form $\mu(\mathbf{e}^4 \wedge \mathbf{f}, \mathbf{e}^4 \wedge \mathbf{f})$ is negative definite. Then $\mathbf{M}$ corresponds to the duality operator $\mathbf{D}_2$ associated with some pair $(\sigma, \mathbf{\Xi})$ only if $\mathbf{\Xi}$ is in the same direction as $\bar{\mathbf{\Xi}}$. Since $\mathbf{D}_2$ is invariant with respect to any change of gauge, without loss of generality we may assume that $\Xi = \bar{\Xi}$. Then the correspondence between **M** and σ is one-to-one according to the uniqueness theorem proved in the preceding section.

It turns out that the three basic properties (M-1), (M-2), and (M-3) are not only necessary but also sufficient for the existence of a pair (σ, Ξ) such that (52.3) holds. To prove sufficiency, we need certain preliminary lemmas, which we shall summarize in this section. These lemmas as well as the existence theorem in the following section are given by Toupin[(3)] in his lectures.

Lemma 1. A tensor **X** in $\mathscr{E}_{\mathbf{w}}^* \wedge \mathscr{E}_{\mathbf{w}}^*$ is simple; i.e.,

$$\mathbf{X} = \mathbf{e}^1 \wedge \mathbf{e}^2 \tag{53.12}$$

for some $\mathbf{e}^1$ and $\mathbf{e}^2$ in $\mathscr{E}_{\mathbf{w}}^*$, if and only if

$$\langle \boldsymbol{\Delta}_2(\mathbf{X}), \mathbf{X} \rangle = 0. \tag{53.13}$$

Proof. Necessity. When $\mathbf{X} = \mathbf{0}$, (53.13) certainly holds. When $\mathbf{X} \neq \mathbf{0}$, the covariant vectors $\mathbf{e}^1$ and $\mathbf{e}^2$ must be linearly independent. Then we can extend the set $\{\mathbf{e}^1, \mathbf{e}^2\}$ into a positive unit basis $\{\mathbf{e}^\alpha\}$ with respect to Ξ. From (50.10a)

$$\langle \boldsymbol{\Delta}_2(\mathbf{e}^1 \wedge \mathbf{e}^2), \mathbf{e}^1 \wedge \mathbf{e}^2 \rangle = \langle \mathbf{e}_3 \wedge \mathbf{e}_4, \mathbf{e}^1 \wedge \mathbf{e}^2 \rangle = 0. \tag{53.14}$$

Thus (53.13) is necessary for **X** to be simple.

Sufficiency. Let **X** be given by a component form in general relative to a positive unit basis $\{\mathbf{e}^\alpha\}$, viz.,

$$\mathbf{X} = \alpha^1\mathbf{e}^1 \wedge \mathbf{e}^2 + \alpha^2\mathbf{e}^3 \wedge \mathbf{e}^4 + \alpha^3\mathbf{e}^2 \wedge \mathbf{e}^3 + \alpha^4\mathbf{e}^1 \wedge \mathbf{e}^4 + \alpha^5\mathbf{e}^2 \wedge \mathbf{e}^4 + \alpha^6\mathbf{e}^3 \wedge \mathbf{e}^1. \tag{53.15}$$

Then from (50.10) and (53.13)

$$\langle \boldsymbol{\Delta}_2(\mathbf{X}), \mathbf{X} \rangle = 2(\alpha^1\alpha^2 + \alpha^3\alpha^4 + \alpha^5\alpha^6) = 0. \tag{53.16}$$

From (53.15) the tensor component matrix $[X_{\mu\nu}]$ of **X** relative to $\{\mathbf{e}^\alpha\}$ is

$$[X_{\mu\nu}] = \begin{bmatrix} 0 & \alpha^1 & -\alpha^6 & \alpha^4 \\ -\alpha^1 & 0 & \alpha^3 & \alpha^5 \\ \alpha^6 & -\alpha^3 & 0 & \alpha^2 \\ -\alpha^4 & -\alpha^5 & -\alpha^2 & 0 \end{bmatrix}. \tag{53.17}$$

(3) R. A. Toupin, "Elasticity and electro-magnetics" (see footnote on page 351).

In particular,

$$\det[X_{\mu\nu}] = (\alpha^1\alpha^2 + \alpha^3\alpha^4 + \alpha^5\alpha^6)^2 = 0. \tag{53.18}$$

From (53.18) we may choose one of the basis vectors of the dual basis $\{\mathbf{e}_\alpha\}$, say, $\mathbf{e}_4$, in such a way that

$$\mathbf{X}(\mathbf{e}_4) = \mathbf{0}. \tag{53.19}$$

Relative to such a basis $\{\mathbf{e}^4\}$ the component form of $\mathbf{X}$ reduces to

$$\mathbf{X} = \alpha^1\mathbf{e}^1 \wedge \mathbf{e}^2 + \alpha^3\mathbf{e}^2 \wedge \mathbf{e}^3 + \alpha^6\mathbf{e}^3 \wedge \mathbf{e}^1, \tag{53.20}$$

and the tensor component matrix $[X_{\mu\nu}]$ reduces to

$$[X_{\mu\nu}] = \begin{bmatrix} 0 & \alpha^1 & -\alpha^6 & 0 \\ -\alpha^1 & 0 & \alpha^3 & 0 \\ \alpha^6 & -\alpha^3 & 0 & 0 \\ 0 & 0 & 0 & 0 \end{bmatrix}. \tag{53.21}$$

Now since the determinant of a 3×3 skew-symmetric matrix is always equal to zero, by the same argument as that from (53.18) to (53.19) we may choose $\mathbf{e}_3$ in such a way that

$$\mathbf{X}(\mathbf{e}_3) = \mathbf{0}. \tag{53.22}$$

Then the component form of $\mathbf{X}$ reduces to

$$\mathbf{X} = \alpha^1\mathbf{e}^1 \wedge \mathbf{e}^2. \tag{53.23}$$

Thus (53.13) is sufficient for $\mathbf{X}$ to be simple. □

Lemma 2. Two simple tensors $\mathbf{X}$ and $\mathbf{Y}$ in $\mathscr{E}_{\mathbf{w}}{}^* \wedge \mathscr{E}_{\mathbf{w}}{}^*$ share a common factor, i.e.,

$$\mathbf{Y} = \mathbf{e}^1 \wedge \mathbf{e}^2, \qquad \mathbf{Y} = \mathbf{e}^1 \wedge \mathbf{e}^3 \tag{53.24}$$

for some $\mathbf{e}^1$, $\mathbf{e}^2$, and $\mathbf{e}^3$ in $\mathscr{E}_{\mathbf{w}}{}^*$, if and only if

$$\iota(\mathbf{X}, \mathbf{Y}) \equiv \langle \mathbf{\Delta}_2(\mathbf{X}), \mathbf{Y} \rangle = 0. \tag{53.25}$$

Note. The bilinear form

$$\iota\colon \mathscr{E}_{\mathbf{w}}{}^* \wedge \mathscr{E}_{\mathbf{w}}{}^* \times \mathscr{E}_{\mathbf{w}}{}^* \wedge \mathscr{E}_{\mathbf{w}}{}^* \to \mathscr{R} \tag{53.26}$$

defined by (53.25a) corresponds to that of the identity map $\mathbf{I}_2$ on $\mathscr{E}_{\mathbf{w}}{}^* \wedge \mathscr{E}_{\mathbf{w}}{}^*$ in accord with the general definition (53.5). (Here $\bar{\mathbf{\Xi}} = \mathbf{\Xi}$.)

Proof. When **X** and **Y** are proportional, the assertion of this lemma reduces to that of the preceding lemma. Hence it suffices to prove the lemma for a linearly independent pair $\{\mathbf{X}, \mathbf{Y}\}$.

Necessity. When **X** and **Y** are linearly independent and have the forms shown in (53.24), the set $\{\mathbf{e}^1, \mathbf{e}^2, \mathbf{e}^3\}$ must be linear independent. Hence we can extend the set $\{\mathbf{e}^1, \mathbf{e}^2, \mathbf{e}^3\}$ into a positive unit basis $\{\mathbf{e}^\alpha\}$. From (50.10) and (53.24)

$$\iota(\mathbf{X}, \mathbf{Y}) = \langle \mathbf{e}_3 \wedge \mathbf{e}_4, \mathbf{e}^1 \wedge \mathbf{e}^3 \rangle = 0. \tag{53.27}$$

Thus (53.25) is necessary for (53.24).

Sufficiency. Suppose that **X** and **Y** do not share any common factor. Then they are of the forms

$$\mathbf{Y} = \mathbf{e}^1 \wedge \mathbf{e}^2, \qquad \mathbf{Y} = \mathbf{e}^3 \wedge \mathbf{e}^4 \tag{53.28}$$

such that no nonzero linear combination of $\{\mathbf{e}^1, \mathbf{e}^2\}$ is equal to a linear combination of $\{\mathbf{e}^3, \mathbf{e}^4\}$. This condition means precisely that $\{\mathbf{e}_\alpha\}$ is a basis for $\mathscr{E}_\mathbf{w}{}^*$. In general, $\{\mathbf{e}^\alpha\}$ need not be a positive unit basis with respect to $\boldsymbol{\Xi}$, but clearly it may be transformed into one by replacing one of its elements, say $\mathbf{e}^1$, by $a\mathbf{e}^1$, where a is an appropriate nonzero number. Hence if we replace (53.24) by

$$\mathbf{X} = \frac{1}{a}\, \mathbf{e}^1 \wedge \mathbf{e}^2, \qquad \mathbf{Y} = \mathbf{e}^3 \wedge \mathbf{e}^4, \tag{53.29}$$

then we may assume that $\{\mathbf{e}^\alpha\}$ is a positive unit basis with respect to $\boldsymbol{\Xi}$. From (50.10) and (53.29)

$$\iota(\mathbf{X}, \mathbf{Y}) = \left\langle \frac{1}{a}\, \mathbf{e}_3 \wedge \mathbf{e}_4, \mathbf{e}^3 \wedge \mathbf{e}^4 \right\rangle = \frac{1}{a} \neq 0. \tag{53.30}$$

Thus (53.25) is sufficient for (53.24). □

Lemma 3. If **X** is a simple tensor in $\mathscr{E}_\mathbf{w}{}^* \wedge \mathscr{E}_\mathbf{w}{}^*$, then $\mathbf{M}(\mathbf{X})$ is also.

Proof. We verify the condition (53.13) for $\mathbf{M}(\mathbf{X})$, viz.,

$$\begin{aligned}\iota(\mathbf{M}(\mathbf{X}), \mathbf{M}(\mathbf{X})) &= \langle \boldsymbol{\Delta}_2\mathbf{M}(\mathbf{X}), \mathbf{M}(\mathbf{X})\rangle = \mu(\mathbf{M}(\mathbf{X}), \mathbf{X}) = \mu(\mathbf{X}, \mathbf{M}(\mathbf{X}))\\ &= \langle \boldsymbol{\Delta}_2(\mathbf{X}), \mathbf{M}(\mathbf{M}(\mathbf{X}))\rangle = \langle \boldsymbol{\Delta}_2(\mathbf{X}), -\mathbf{X}\rangle \\ &= -\iota(\mathbf{X}, \mathbf{X}) = 0,\end{aligned} \tag{53.31}$$

where we have used the basic properties (M-1) and (M-2). □

Lemma 4. If $\mathbf{X}$ and $\mathbf{Y}$ are simple tensors in $\mathscr{E}_\mathbf{w}^* \wedge \mathscr{E}_\mathbf{w}^*$ sharing a common factor, then $\mathbf{M}(\mathbf{X})$ and $\mathbf{M}(\mathbf{Y})$ are also.

Proof. We verify the condition (53.25) for $\mathbf{M}(\mathbf{X})$ and $\mathbf{M}(\mathbf{Y})$, viz.,

$$\begin{aligned}\iota(\mathbf{M}(\mathbf{X}), \mathbf{M}(\mathbf{Y})) &= \langle \boldsymbol{\Delta}_2\mathbf{M}(\mathbf{X}), \mathbf{M}(\mathbf{Y})\rangle = \mu(\mathbf{M}(\mathbf{X}), \mathbf{Y}) = \mu(\mathbf{Y}, \mathbf{M}(\mathbf{X}))\\ &= \langle \boldsymbol{\Delta}_2(\mathbf{Y}), \mathbf{M}(\mathbf{M}(\mathbf{X}))\rangle = \langle \boldsymbol{\Delta}_2(\mathbf{Y}), -\mathbf{X}\rangle\\ &= -\iota(\mathbf{X}, \mathbf{Y}) = 0,\end{aligned} \tag{53.32}$$

where we have used the basic properties (M-1) and (M-2) again. □

Lemma 5. For any nonzero tensor $\mathbf{X}$ in $\mathscr{E}_\mathbf{w}^* \wedge \mathscr{E}_\mathbf{w}^*$ the tensors $\mathbf{M}(\mathbf{X})$ and $\mathbf{X}$ must be linearly independent.

Proof. Suppose that $\mathbf{M}(\mathbf{X})$ is proportional to $\mathbf{X}$, say,

$$\mathbf{M}(\mathbf{X}) = a\mathbf{X}. \tag{53.33}$$

Applying $\mathbf{M}$ to this equation and using (53.1), we get

$$\mathbf{M}^2(\mathbf{X}) = -\mathbf{X} = a^2\mathbf{X}. \tag{53.34}$$

Thus

$$(1 + a^2)\mathbf{X} = \mathbf{0}, \tag{53.35}$$

which contradicts the hypothesis that $\mathbf{X}$ is nonzero. □

In the following section we shall use the preceding five lemmas to show that when $\mathbf{M}$ possesses the basic properties (M-1), (M-2), and (M-3), there is a pair $(\sigma, \boldsymbol{\Xi})$ such that the condition (52.3) holds. In the proof of this existence theorem many simple tensors in $\mathscr{E}_\mathbf{w}^* \wedge \mathscr{E}_\mathbf{w}^*$ shall be used, so we now make an important remark regarding them: Any nonzero simple tensor in $\mathscr{E}_\mathbf{w}^* \wedge \mathscr{E}_\mathbf{w}^*$, such as the simple tensor $\mathbf{X}$ given explicitly by (53.12), corresponds to a (2-dimensional) volume tensor on the plane in $\mathscr{E}_\mathbf{w}^*$ spanned by the factors; i.e., the plane span$\{\mathbf{e}^1, \mathbf{e}^2\}$. This remark is discussed in detail in Section 40 of IVT-1. By virtue of that remark the simple tensor $\mathbf{X}$ may be represented not only by (53.12) but also by

$$\mathbf{X} = (a\mathbf{e}^1 + b\mathbf{e}^2) \wedge (c\mathbf{e}^1 + d\mathbf{e}^2), \tag{53.36}$$

provided that

$$\det\begin{bmatrix} a & b \\ c & d \end{bmatrix} = ad - bc = 1. \tag{53.37}$$

Consequently, any nonzero linear combination $a\mathbf{e}^1 + b\mathbf{e}^2 \neq \mathbf{0}$ is a factor of $\mathbf{X}$. If one factor $a\mathbf{e}^1 + b\mathbf{e}^2$ of $\mathbf{X}$ is chosen, then the other factor may be any nonzero linear combination $c\mathbf{e}^1 + d\mathbf{e}^2$, which is linearly independent of $a\mathbf{e}^1 + b\mathbf{e}^2$ and is normalized to satisfy the condition (53.37). Since the possible values of (a, b, c, d) form a 3-dimensional surface characterized by the algebraic equation (53.37), there are many factors of $\mathbf{X}$ differing from the factors $\mathbf{e}^1$ and $\mathbf{e}^2$ shown in (53.12).

In Toupin's original lecture notes he seemed to have overlooked this simple fact. For instance in his proof of the existence theorem he argued that[(4)] if a covariant vector $\mathbf{f}$ is known to be a factor of the simple tensor $\mathbf{X} = \mathbf{e}^1 \wedge \mathbf{e}^2$, and if $\mathbf{f}$ is shown to be not proportional to $\mathbf{e}^1$, then $\mathbf{f}$ must be proportional to $\mathbf{e}^2$. Such an argument leaves a gap in the proof, since there are many factors of $\mathbf{X}$ not proportional to either $\mathbf{e}^1$ or $\mathbf{e}^2$.

In our proof of sufficiency in Lemma 2 we have noted that the simple tensors $\mathbf{X}$ and $\mathbf{Y}$ given by (53.28) share no common factor if and only if no nonzero linear combination of $\{\mathbf{e}^1, \mathbf{e}^2\}$ coincides with a linear combination of $\{\mathbf{e}^3, \mathbf{e}^4\}$, or, equivalently,

$$\operatorname{span}\{\mathbf{e}^1, \mathbf{e}^2\} \cap \operatorname{span}\{\mathbf{e}^3, \mathbf{e}^4\} = \{\mathbf{0}\}, \tag{53.38}$$

which is necessary and sufficient for $\{\mathbf{e}^1, \mathbf{e}^2, \mathbf{e}^3, \mathbf{e}^4\}$ to be a basis of $\mathscr{E}_{\mathbf{w}}{}^*$. That assertion reflects clearly that a factor of $\mathbf{X}$ is not just $\mathbf{e}^1$ or $\mathbf{e}^2$ but may be any nonzero linear combination of $\{\mathbf{e}^1, \mathbf{e}^2\}$. By the same token a factor of $\mathbf{Y}$ is not just $\mathbf{e}^3$ or $\mathbf{e}^4$ but may be any nonzero linear combination of $\{\mathbf{e}^3, \mathbf{e}^4\}$. It is quite possible for $\mathbf{X}$ and $\mathbf{Y}$ to share a common factor when $\mathbf{e}^1$, $\mathbf{e}^2$, $\mathbf{e}^3$, and $\mathbf{e}^4$ are completely distinct covariant vectors in $\mathscr{E}_{\mathbf{w}}{}^*$.

54. The Maxwell–Lorentz Ether Relation and the Minkowskian Metric III: Toupin's Existence Theorem

In the preceding section we showed that the Maxwell–Lorentz ether tensor $\mathbf{M}(\mathbf{w})$ must possess three basic properties, namely, reciprocity, symmetry, and definiteness, in order that $\mathbf{M}(\mathbf{w})$ may be the same as the duality operator $\mathbf{D}_2(\mathbf{w})$ associated with some pair $(\sigma(\mathbf{w}), \boldsymbol{\Xi}(\mathbf{w}))$. We have remarked that, by using the basic property (M-3), we may choose in a definite way a certain nonzero tensor in $\mathscr{E}_{\mathbf{w}}{}^* \wedge \mathscr{E}_{\mathbf{w}}{}^* \wedge \mathscr{E}_{\mathbf{w}}{}^* \wedge \mathscr{E}_{\mathbf{w}}{}^*$ to be the positive unit density tensor $\boldsymbol{\Xi}(\mathbf{w})$ in the pair $(\sigma(\mathbf{w}), \boldsymbol{\Xi}(\mathbf{w}))$. We now

(4) See, for example, p. 281 in Toupin's lecture notes, "Elasticity and electro-magnetics" (see footnote on page 351).

show that the three properties suffice to determine a certain Minkowskian inner product $\sigma(\mathbf{w})$ on $\mathscr{E}_{\mathbf{w}}$ such that

$$[\sigma_4(\mathbf{w})](\Xi(\mathbf{w}), \Xi(\mathbf{w})) = -1, \tag{54.1a}$$

$$\mathbf{M}(\mathbf{w}) = \mathbf{D}_2(\mathbf{w}), \tag{54.1b}$$

where $\mathbf{D}_2(\mathbf{w})$ denotes the duality operator associated with the pair $(\sigma(\mathbf{w}), \Xi(\mathbf{w}))$. The condition (54.1a) simply means that $\Xi(\mathbf{w})$ is a unit density tensor with respect to $\sigma(\mathbf{w})$, as it should be. The orientation of $\Xi(\mathbf{w})$ is chosen in such a way that the quadratic form $\mu(\mathbf{X}, \mathbf{X})$ mentioned in the basic property (M-3) is negative definite. We state the preceding result formally as a theorem.

Existence Theorem (Toupin[(5)]). Suppose that a linear operator $\mathbf{M}(\mathbf{w})$ of the form (52.1) possesses the three basic properties (M-1), (M-2), and (M-3), and let $\Xi(\mathbf{w})$ be chosen in such a way that the quadratic form $\mu(\mathbf{X}, \mathbf{X})$ in (M-3) is negative definite. Then there is a Minkowskian inner product $\sigma(\mathbf{w})$ on $\mathscr{E}_{\mathbf{w}}$ such that (54.1) holds.

Like the uniqueness theorem proved in Section 52, the existence theorem is strictly a result for each point $\mathbf{w}$ in $\mathscr{E}$. Hence we shall restrict our attention to tensors at a particular point $\mathbf{w}$. For simplicity of writing we shall now drop from the notation the argument $\mathbf{w}$ as before.

To prove the existence of σ, we use first the basic property (M-3) and choose a particular covariant vector $\mathbf{f}^4$ in $\mathscr{E}_{\mathbf{w}}^*$ such that

$$\mu(\mathbf{X}, \mathbf{X}) < 0 \tag{54.2}$$

for all nonzero simple tensors $\mathbf{X}$ in $\mathscr{E}_{\mathbf{w}}^* \wedge \mathscr{E}_{\mathbf{w}}^*$ having $\mathbf{f}^4$ as a factor. From (54.2) it is clear that $\mathbf{f}^4$ is nonzero. We choose an arbitrary plane in $\mathscr{E}_{\mathbf{w}}^*$ containing $\mathbf{f}^4$ and designate another nonzero covariant vector in that plane, not parallel to $\mathbf{f}^4$, by $\mathbf{f}^1$. Then we put

$$\mathbf{X} = \mathbf{f}^1 \wedge \mathbf{f}^4. \tag{54.3}$$

Of course, this simple tensor is nonzero and satisfies (54.2).

From Lemmas 3 and 5 in the preceding section $\mathbf{M}(\mathbf{X})$ must be a

(5) R. A. Toupin, "Elasticity and electro-magnetics" (see footnote on page 351). Toupin did not state his result in the form presented here, but his result may be reformulated and converted into the theorem here.

nonzero simple tensor which is not proportional to $\mathbf{X}$. We claim that $\mathbf{X}$ and $\mathbf{M}(\mathbf{X})$ cannot even share a common factor. This assertion may be verified easily by using the criterion given by Lemma 2, viz.,

$$\iota(\mathbf{X}, \mathbf{M}(\mathbf{X})) = \langle \mathbf{\Delta}_2(\mathbf{X}), \mathbf{M}(\mathbf{X}) \rangle = \mu(\mathbf{X}, \mathbf{X}) < 0. \tag{54.4}$$

Hence $\mathbf{M}(\mathbf{X})$ may be expressed as

$$\mathbf{M}(\mathbf{X}) = \mathbf{f}^3 \wedge \mathbf{f}^2, \tag{54.5}$$

and the set $\{\mathbf{f}^\alpha\}$ forms a basis for $\mathscr{E}_{\mathbf{w}}{}^*$.

Now we define another nonzero simple tensor $\mathbf{Y}$ by

$$\mathbf{Y} = \mathbf{f}^2 \wedge \mathbf{f}^4, \tag{54.6}$$

which also has $\mathbf{f}^4$ as a factor. Hence from (M-3)

$$\mu(\mathbf{Y}, \mathbf{Y}) < 0. \tag{54.7}$$

By the argument as before $\mathbf{M}(\mathbf{Y})$ must be a nonzero simple tensor, and $\mathbf{Y}$ and $\mathbf{M}(\mathbf{Y})$ do not share a common factor. In particular, both $\mathbf{f}^2$ and $\mathbf{f}^4$ are not factors of $\mathbf{M}(\mathbf{Y})$.

From (54.6) and (54.3) we see that $\mathbf{X}$ and $\mathbf{Y}$ share a common factor, namely, $\mathbf{f}^4$. Hence by Lemma 4, $\mathbf{M}(\mathbf{X})$ and $\mathbf{M}(\mathbf{Y})$ must also share a common factor, say, $\mathbf{g}^3$. From (54.5) this common factor $\mathbf{g}^3$ is a certain nonzero linear combination of $\mathbf{f}^3$ and $\mathbf{f}^2$, say,

$$\mathbf{g}^3 = a\mathbf{f}^3 + b\mathbf{f}^2, \tag{54.8}$$

where the coefficient a is nonzero since $\mathbf{f}^2$ is not a factor of $\mathbf{M}(\mathbf{Y})$. Since a common factor of $\mathbf{M}(\mathbf{X})$ and $\mathbf{M}(\mathbf{Y})$ is unique to within an arbitrary nonzero multiple only, the nonzero coefficient a in (54.8) is arbitrary, but for each choice of a there is a unique corresponding b, such that the covariant vector $\mathbf{g}^3$ given by (54.8) is a common factor of $\mathbf{M}(\mathbf{X})$ and $\mathbf{M}(\mathbf{Y})$.

Now because the coefficient a in (54.8) is nonzero, if we put

$$\mathbf{g}^1 = \mathbf{f}^1, \qquad \mathbf{g}^2 = \mathbf{f}^2, \qquad \mathbf{g}^4 = \mathbf{f}^4, \tag{54.9}$$

then $\{\mathbf{g}^\alpha\}$ is a basis for $\mathscr{E}_{\mathbf{w}}{}^*$. From (54.8) the density tensors associated with the bases $\{\mathbf{f}^\alpha\}$ and $\{\mathbf{g}^\alpha\}$ are related by

$$\mathbf{g}^1 \wedge \mathbf{g}^2 \wedge \mathbf{g}^3 \wedge \mathbf{g}^4 = a\mathbf{f}^1 \wedge \mathbf{f}^2 \wedge \mathbf{f}^3 \wedge \mathbf{f}^4. \tag{54.10}$$

This relation implies that we can choose an appropriate nonzero number a

such that $\{\mathbf{g}^\alpha\}$ is a positive unit basis relative to the density tensor $\boldsymbol{\Xi}$; i.e.,

$$\mathbf{g}^1 \wedge \mathbf{g}^2 \wedge \mathbf{g}^3 \wedge \mathbf{g}^4 = \boldsymbol{\Xi}. \tag{54.11}$$

Using the preceding basis $\{\mathbf{g}^\alpha\}$, we rewrite the equations (54.3), (54.6), and (54.5) as

$$\mathbf{X} = \mathbf{g}^1 \wedge \mathbf{g}^4, \tag{54.12a}$$

$$\mathbf{Y} = \mathbf{g}^2 \wedge \mathbf{g}^4, \tag{54.12b}$$

$$\mathbf{M}(\mathbf{X}) = \frac{1}{a}\, \mathbf{g}^3 \wedge \mathbf{g}^2, \tag{54.12c}$$

where in (54.12c) we have used the inverse of (54.8), viz.,

$$\mathbf{f}^3 = \frac{1}{a}\,(\mathbf{g}^3 - b\mathbf{g}^2). \tag{54.13}$$

We claim that the nonzero number a, which enters into (54.12c), must be positive. Indeed, since $\{\mathbf{g}^\alpha\}$ is a positive unit basis with respect to $\boldsymbol{\Xi}$, we may use the component formula (50.10d) to calculate $\boldsymbol{\Delta}_2(\mathbf{X})$, viz.,

$$\boldsymbol{\Delta}_2(\mathbf{X}) = \boldsymbol{\Delta}_2(\mathbf{g}^1 \wedge \mathbf{g}^4) = -\mathbf{g}_3 \wedge \mathbf{g}_2. \tag{54.14}$$

Then from (54.4), (54.12), and (54.2),

$$\begin{aligned}\iota(\mathbf{X}, \mathbf{M}(\mathbf{X})) &= \left\langle \boldsymbol{\Delta}_2(\mathbf{g}^1 \wedge \mathbf{g}^4), \frac{1}{a}\, \mathbf{g}^3 \wedge \mathbf{g}^2 \right\rangle = \left\langle -\mathbf{g}_3 \wedge \mathbf{g}_2, \frac{1}{a}\, \mathbf{g}^3 \wedge \mathbf{g}^2 \right\rangle = -\frac{1}{a} \\ &= \mu(\mathbf{X}, \mathbf{X}) < 0.\end{aligned} \tag{54.15}$$

Next we claim that $\mathbf{X}$ and $\mathbf{M}(\mathbf{Y})$ share a common factor. As before we verify this assertion by using the criterion given by Lemma 2, viz.,

$$\begin{aligned}\iota(\mathbf{X}, \mathbf{M}(\mathbf{Y})) &= \langle \boldsymbol{\Delta}_2(\mathbf{X}), \mathbf{M}(\mathbf{Y}) \rangle = \mu(\mathbf{X}, \mathbf{Y}) = \mu(\mathbf{Y}, \mathbf{X}) = \langle \boldsymbol{\Delta}_2(\mathbf{Y}), \mathbf{M}(\mathbf{X}) \rangle \\ &= \left\langle \boldsymbol{\Delta}_2(\mathbf{g}^2 \wedge \mathbf{g}^4), \frac{1}{a}\, \mathbf{g}^3 \wedge \mathbf{g}^2 \right\rangle = \frac{1}{a} \langle -\mathbf{g}_1 \wedge \mathbf{g}_3, \mathbf{g}^3 \wedge \mathbf{g}^2 \rangle = 0,\end{aligned} \tag{54.16}$$

where we have used (54.12), (50.10e), and the basic property (M-2). From (54.12a) the common factor of $\mathbf{X}$ and $\mathbf{M}(\mathbf{Y})$ is a certain nonzero linear combination of $\mathbf{g}^1$ and $\mathbf{g}^4$, say,

$$\mathbf{h}^1 = \bar{a}\mathbf{g}^1 + \bar{b}\mathbf{g}^4, \tag{54.17}$$

where the coefficient $\bar{a}$ is nonzero since we have remarked after (54.7) that $\mathbf{f}^4 = \mathbf{g}^4$ is not a factor of $\mathbf{M}(\mathbf{Y})$. Then as explained before, the nonzero coefficient $\bar{a}$ in (54.17) is arbitrary, but for each choice of $\bar{a}$ there is a unique corresponding $\bar{b}$ such that the covariant vector $\mathbf{h}^1$ given by (54.17) is a common factor of $\mathbf{X}$ and $\mathbf{M}(\mathbf{Y})$.

We choose the coefficient $\bar{a}$ to be 1, and we put

$$\mathbf{h}^2 = \mathbf{g}^2, \qquad \mathbf{h}^3 = \mathbf{g}^3, \qquad \mathbf{h}^4 = \mathbf{g}^4. \tag{54.18}$$

Then $\{\mathbf{h}^\alpha\}$, like $\{\mathbf{g}^\alpha\}$, is a positive unit basis for $\mathscr{E}_{\mathbf{w}}{}^*$ with respect to $\mathbf{\Xi}$, viz.,

$$\mathbf{h}^1 \wedge \mathbf{h}^2 \wedge \mathbf{h}^3 \wedge \mathbf{h}^4 = \mathbf{g}^1 \wedge \mathbf{g}^2 \wedge \mathbf{g}^3 \wedge \mathbf{g}^4 = \mathbf{\Xi}. \tag{54.19}$$

From (54.17) with $\bar{a} = 1$, (54.18), and (54.12) we now have

$$\mathbf{X} = \mathbf{h}^1 \wedge \mathbf{h}^4, \tag{54.20a}$$

$$\mathbf{M}(\mathbf{X}) = c_1\mathbf{h}^3 \wedge \mathbf{h}^2, \tag{54.20b}$$

$$\mathbf{Y} = \mathbf{h}^2 \wedge \mathbf{h}^4, \tag{54.20c}$$

$$\mathbf{M}(\mathbf{Y}) = c_2\mathbf{h}^1 \wedge \mathbf{h}^3, \tag{54.20d}$$

where we have replaced the notation for the positive number $1/a$ by c_1, and where $\mathbf{M}(\mathbf{Y})$ is of the form (54.20d) because $\mathbf{h}^1$ and $\mathbf{h}^3 = \mathbf{g}^3$ are linearly independent factors of $\mathbf{M}(\mathbf{Y})$. The nonzero number c_2 in (54.20d), like the number c_1 in (54.20b), is positive. Indeed, from (54.4), (54.20), (50.10e), and (54.7),

$$\begin{aligned}\iota(\mathbf{Y}, \mathbf{M}(\mathbf{Y})) &= \langle \mathbf{\Delta}_2(\mathbf{h}^2 \wedge \mathbf{h}^4), c_2\mathbf{h}^1 \wedge \mathbf{h}^3 \rangle = \langle -\mathbf{h}_1 \wedge \mathbf{h}_3, c_2\mathbf{h}^1 \wedge \mathbf{h}^3 \rangle = -c_2 \\ &= \mu(\mathbf{Y}, \mathbf{Y}) < 0.\end{aligned} \tag{54.21}$$

Next we put

$$\mathbf{Z} = \mathbf{h}^3 \wedge \mathbf{h}^4. \tag{54.22}$$

We claim that $\mathbf{M}(\mathbf{Z})$ must be of the form

$$\mathbf{M}(\mathbf{Z}) = c_3\mathbf{h}^2 \wedge \mathbf{h}^1, \tag{54.23}$$

where c_3 is a certain positive number. To prove this result, we notice first that since $\mathbf{h}^4 = \mathbf{g}^4 = \mathbf{f}^4$ is a factor of $\mathbf{Z}$, from (M-3)

$$\mu(\mathbf{Z}, \mathbf{Z}) < 0. \tag{54.24}$$

Then as before $\mathbf{Z}$ and $\mathbf{M}(\mathbf{Z})$ do not share any common factor. In particular, from (54.22) $\mathbf{h}^3$ and $\mathbf{h}^4$ are not factors of $\mathbf{M}(\mathbf{Z})$.

Now since $\mathbf{Y}$ and $\mathbf{Z}$ share a common factor, namely, $\mathbf{h}^4$, from Lemma 4

$\mathbf{M}(\mathbf{Y})$ and $\mathbf{M}(\mathbf{Z})$ must also. From (54.20d) the common factor of $\mathbf{M}(\mathbf{Y})$ and $\mathbf{M}(\mathbf{Z})$ is a certain nonzero linear combination of $\mathbf{h}^1$ and $\mathbf{h}^3$, say,

$$\mathbf{k}^1 = a_1\mathbf{h}^1 + b_1\mathbf{h}^3, \tag{54.25}$$

where the coefficient a_1 is nonzero since $\mathbf{h}^3$ is not a factor of $\mathbf{M}(\mathbf{Z})$. As before a_1 is arbitrary, but for each choice of a_1 there is a unique corresponding b_1 such that the covariant vector $\mathbf{k}^1$ given by (54.25) is a common factor of $\mathbf{M}(\mathbf{Y})$ and $\mathbf{M}(\mathbf{Z})$.

By the same token $\mathbf{M}(\mathbf{X})$ and $\mathbf{M}(\mathbf{Z})$ must share a common factor, since $\mathbf{X}$ and $\mathbf{Z}$ both have the factor $\mathbf{h}^4$. From (54.20b) the common factor of $\mathbf{M}(\mathbf{X})$ and $\mathbf{M}(\mathbf{Z})$ is a certain nonzero linear combination of $\mathbf{h}^2$ and $\mathbf{h}^3$, say,

$$\mathbf{k}^2 = a_2\mathbf{h}^2 + b_2\mathbf{h}^3, \tag{54.26}$$

where the coefficient a_2 is nonzero. As before, a_2 is arbitrary, but for each choice of a_2 there is a unique corresponding b_2. Because both a_1 and a_2 are nonzero and arbitrary, $\mathbf{k}^1$ and $\mathbf{k}^2$ are linearly independent factors of $\mathbf{M}(\mathbf{Z})$. Hence $\mathbf{M}(\mathbf{Z})$ may be expressed as

$$\mathbf{M}(\mathbf{Z}) = \mathbf{k}^2 \wedge \mathbf{k}^1 = (a_2\mathbf{h}^2 + b_2\mathbf{h}^3) \wedge (a_1\mathbf{h}^1 + b_1\mathbf{h}^3), \tag{54.27}$$

where the product a_2a_1 is fixed.

To complete the proof of (54.23), we must show that the coefficients b_1 and b_2 in (54.27) both vanish. The fact that $b_1 = 0$ follows from the condition that $\mathbf{X}$ and $\mathbf{M}(\mathbf{Z})$ share a common factor. As before we use the criterion given by Lemma 4 to verify that condition, viz.,

$$\begin{aligned} \iota(\mathbf{X}, \mathbf{M}(\mathbf{Z})) &= \langle \mathbf{\Delta}_2(\mathbf{X}), \mathbf{M}(\mathbf{Z})\rangle = \mu(\mathbf{X}, \mathbf{Z}) = \mu(\mathbf{Z}, \mathbf{X}) = \langle \mathbf{\Delta}_2(\mathbf{Z}), \mathbf{M}(\mathbf{X})\rangle \\ &= \langle \mathbf{\Delta}_2(\mathbf{h}^3 \wedge \mathbf{h}^4), c_1\mathbf{h}^3 \wedge \mathbf{h}^2\rangle = \langle -\mathbf{h}_2 \wedge \mathbf{h}_1, c_1\mathbf{h}^3 \wedge \mathbf{h}^2\rangle = 0, \end{aligned} \tag{54.28}$$

where we have used (54.22), (50.10f), and the basic property (M-2). From (54.20a) the common factor of $\mathbf{X}$ and $\mathbf{M}(\mathbf{Z})$ is a certain nonzero linear combination of $\mathbf{h}^1$ and $\mathbf{h}^4$, say,

$$\mathbf{k}^3 = a_3\mathbf{h}^1 + b_3\mathbf{h}^4, \tag{54.29}$$

where the coefficient a_3 is nonzero since $\mathbf{h}^4$ is not a factor of $\mathbf{M}(\mathbf{Z})$. From (54.27) the common factor of $\mathbf{X}$ and $\mathbf{M}(\mathbf{Z})$ is also a certain nonzero linear combination of $\mathbf{k}^1$ and $\mathbf{k}^2$. Thus we have an equation of the form

$$a_4(a_1\mathbf{h}^1 + b_1\mathbf{h}^3) + b_4(a_2\mathbf{h}^2 + b_2\mathbf{h}^3) = a_3\mathbf{h}^1 + b_3\mathbf{h}^4, \tag{54.30}$$

where the coefficients a_1, a_2, and a_3 are nonzero. Since $\{\mathbf{h}^\alpha\}$ is a basis, the equation (54.30) implies that

$$a_1 a_4 - a_3 = 0, \tag{54.31a}$$

$$a_2 b_4 = 0, \tag{54.31b}$$

$$a_4 b_1 + b_4 b_2 = 0, \tag{54.31c}$$

$$b_3 = 0. \tag{54.31d}$$

From (54.31b) $b_4 = 0$, since $a_2 \neq 0$. Then from (54.31a) $a_4 \neq 0$, since a_1 and a_3 are nonzero. It then follows from (54.31c) that $b_1 = 0$, since $a_4 \neq 0$ and $b_4 = 0$.

By the same argument but based on the simple tensor $\mathbf{Y}$ instead of the simple tensor $\mathbf{X}$, we obtain $b_2 = 0$. Thus we have shown that $\mathbf{M}(\mathbf{Z})$ has the form given by (54.23). The fact that c_3, like c_1 and c_2, is a positive number may be verified easily as before. Specifically, from (54.4), (54.22), (54.23), and (54.24),

$$\begin{aligned} \iota(\mathbf{Z}, \mathbf{M}(\mathbf{Z})) &= \langle \mathbf{\Delta}_2(\mathbf{h}^3 \wedge \mathbf{h}^4), c_3\mathbf{h}^2 \wedge \mathbf{h}^1 \rangle = \langle -\mathbf{h}_2 \wedge \mathbf{h}_1, c_3\mathbf{h}^2 \wedge \mathbf{h}^1 \rangle = -c_3 \\ &= \mu(\mathbf{Z}, \mathbf{Z}) < 0. \end{aligned} \tag{54.32}$$

Thus the proof of (54.23) is complete.

Summarizing the result obtained so far, we see that there is a positive unit basis $\{\mathbf{h}^\alpha\}$ in $\mathscr{E}_{\mathbf{w}}{}^*$ with respect to $\mathbf{\Xi}$ such that

$$\mathbf{X} = \mathbf{h}^1 \wedge \mathbf{h}^4, \tag{54.33a}$$

$$\mathbf{M}(\mathbf{X}) = c_1\mathbf{h}^3 \wedge \mathbf{h}^2, \tag{54.33b}$$

$$\mathbf{Y} = \mathbf{h}^2 \wedge \mathbf{h}^4, \tag{54.33c}$$

$$\mathbf{M}(\mathbf{Y}) = c_2\mathbf{h}^1 \wedge \mathbf{h}^3, \tag{54.33d}$$

$$\mathbf{Z} = \mathbf{h}^3 \wedge \mathbf{h}^4, \tag{54.33e}$$

$$\mathbf{M}(\mathbf{Z}) = c_3\mathbf{h}^2 \wedge \mathbf{h}^1, \tag{54.33f}$$

where c_1, c_2, and c_3 are certain positive numbers. Now we make a final adjustment on the basis by replacing $\mathbf{h}^1$, $\mathbf{h}^2$, $\mathbf{h}^3$, and $\mathbf{h}^4$ by

$$\mathbf{e}^1 = \alpha_1\mathbf{h}^1, \qquad \mathbf{e}^2 = \alpha_2\mathbf{h}^2, \qquad \mathbf{e}^3 = \alpha_3\mathbf{h}^3, \qquad \mathbf{e}^4 = \alpha_4\mathbf{h}^4, \tag{54.34}$$

where the coefficients α_1, α_2, α_3, and α_4 are chosen in such a way that

$$\alpha_1\alpha_2\alpha_3\alpha_4 = 1, \tag{54.35a}$$

$$\alpha_3\alpha_2 = c_1\alpha_1\alpha_4, \tag{54.35b}$$

$$\alpha_1\alpha_3 = c_2\alpha_2\alpha_4, \tag{54.35c}$$

$$\alpha_2\alpha_1 = c_3\alpha_3\alpha_4. \tag{54.35d}$$

The general solution of (54.35) is given by

$$\begin{aligned} \alpha_1 &= \pm(c_2c_3/c_1)^{1/4}, \\ \alpha_2 &= \pm(c_1c_3/c_2)^{1/4}, \\ \alpha_3 &= \pm(c_1c_2/c_3)^{1/4}, \\ \alpha_4 &= \pm(1/c_1c_2c_3)^{1/4}, \end{aligned} \tag{54.36}$$

where any even number of positive sign and even number of negative sign may be used in (54.36), e.g., $\alpha_1 < 0$, $\alpha_2 > 0$, $\alpha_3 > 0$, $\alpha_4 < 0$, etc.

After this adjustment is made, the new basis $\{\mathbf{e}^\alpha\}$ is still a positive unit basis with respect to Ξ, viz.,

$$\mathbf{e}^1 \wedge \mathbf{e}^2 \wedge \mathbf{e}^3 \wedge \mathbf{e}^4 = \alpha_1\alpha_2\alpha_3\alpha_4\mathbf{h}^1 \wedge \mathbf{h}^2 \wedge \mathbf{h}^3 \wedge \mathbf{h}^4 = \Xi, \tag{54.37}$$

where we have used (54.35*a*) and (54.19). In terms of the new basis $\{\mathbf{e}^\alpha\}$ the previous conditions (54.33b), (54.33d), and (54.33f) may be rewritten as

$$\begin{aligned} \mathbf{M}(\mathbf{e}^1 \wedge \mathbf{e}^4) &= \alpha_1\alpha_4\mathbf{M}(\mathbf{h}^1 \wedge \mathbf{h}^4) = c_1\alpha_1\alpha_4\mathbf{h}^3 \wedge \mathbf{h}^2 = \frac{c_1\alpha_1\alpha_4}{\alpha_3\alpha_2}\mathbf{e}^3 \wedge \mathbf{e}^2 \\ &= \mathbf{e}^3 \wedge \mathbf{e}^2, \\ \mathbf{M}(\mathbf{e}^2 \wedge \mathbf{e}^4) &= \alpha_2\alpha_4\mathbf{M}(\mathbf{h}^2 \wedge \mathbf{h}^4) = c_2\alpha_2\alpha_4\mathbf{h}^1 \wedge \mathbf{h}^3 = \frac{c_2\alpha_2\alpha_4}{\alpha_1\alpha_3}\mathbf{e}^1 \wedge \mathbf{e}^3 \\ &= \mathbf{e}^1 \wedge \mathbf{e}^3, \\ \mathbf{M}(\mathbf{e}^3 \wedge \mathbf{e}^4) &= \alpha_3\alpha_4\mathbf{M}(\mathbf{h}^3 \wedge \mathbf{h}^4) = c_3\alpha_3\alpha_4\mathbf{h}^2 \wedge \mathbf{h}^1 = \frac{c_3\alpha_3\alpha_4}{\alpha_2\alpha_1}\mathbf{e}^2 \wedge \mathbf{e}^1 \\ &= \mathbf{e}^2 \wedge \mathbf{e}^1, \end{aligned} \tag{54.38}$$

where we have used (54.35b)–(54.35d). Applying the operator **M** to the preceding conditions and using the basic property (M-1), we obtain

$$\begin{aligned} \mathbf{M}(\mathbf{e}^3 \wedge \mathbf{e}^2) &= \mathbf{M}^2(\mathbf{e}^1 \wedge \mathbf{e}^4) = -\mathbf{e}^1 \wedge \mathbf{e}^4, \\ \mathbf{M}(\mathbf{e}^1 \wedge \mathbf{e}^3) &= \mathbf{M}^2(\mathbf{e}^2 \wedge \mathbf{e}^4) = -\mathbf{e}^2 \wedge \mathbf{e}^4, \\ \mathbf{M}(\mathbf{e}^2 \wedge \mathbf{e}^1) &= \mathbf{M}^2(\mathbf{e}^3 \wedge \mathbf{e}^4) = -\mathbf{e}^3 \wedge \mathbf{e}^4. \end{aligned} \tag{54.39}$$

Comparing the conditions (54.38) and (54.39) with the component formula (40.31) for the duality operator $\mathbf{D}_2$, we see that **M** coincides with $\mathbf{D}_2$ provided that $\{\mathbf{e}^\alpha\}$ is regarded as a positive orthonormal basis with respect to a Minkowskian inner product σ_1 on $\mathscr{E}_{\mathbf{w}}{}^*$. Thus the existence theorem is proved. □

Combining the preceding existence theorem with the uniqueness theorem proved in Section 52, we see that a Minkowskian metric and an orientation on $\mathscr{E}$ may be determined to within an arbitrary change of gauge by a field $\mathbf{M}$ which possesses the three basic properties (M-1), (M-2), and (M-3). Conversely, if a Minkowskian metric and an orientation are given on $\mathscr{E}$, then the field $\mathbf{D}_2$ possesses the three basic properties (M-1), (M-2), and (M-3). The preceding result is quite similar to the result obtained in Section 50 on the basis of the field $\mathbf{N} = \mathbf{D}_1$.

55. The Electromagnetic Action and the Electromagnetic Stress-Energy-Momentum Tensor

In the special theory of relativity we have shown that an electromagnetic field $\mathbf{\Phi}$ gives rise to a field $\mathbf{\Omega}$ of electromagnetic stress-energy-momentum tensors. Using the Lorentz electron theory, we can express $\mathbf{\Omega}$ by the component formula

$$\Omega_{\alpha\beta} = \frac{1}{4\pi}\left(\Sigma^{\mu\nu}\Phi_{\mu\alpha}\Phi_{\nu\beta} - \frac{1}{4}\Sigma_{\alpha\beta}\Phi^{\mu\nu}\Phi_{\mu\nu}\right); \tag{55.1}$$

cf. (41.19). We have also shown that $\mathbf{\Omega}$ satisfies the equation of balance

$$\operatorname{Div}\mathbf{\Omega} + \boldsymbol{\xi} = \mathbf{0}; \tag{55.2}$$

cf. (41.18), where the divergence is taken with respect to the curvature-free Minkowskian metric on the Minkowskian space–time. In (55.2) $\boldsymbol{\xi}$ denotes the field of momentum-energy production vectors and is given by the component formula

$$\xi^{\alpha} = \Phi^{\alpha\beta}J_{\beta}; \tag{55.3}$$

cf. (41.8), where $\mathbf{J}$ denotes the dual of the charge–current field $\mathbf{\Psi}$ as shown in (41.5).

Now it turns out that the equation of balance (55.2) remains valid in the context of the general theory of relativity; i.e., if we define $\mathbf{\Omega}$ by (55.1) and $\boldsymbol{\xi}$ by (55.3), then Maxwell's equations together with the Maxwell–Lorentz ether relation imply (55.2), where the divergence is taken with respect to the Minkowskian metric σ on the Minkowskian manifold $\mathscr{E}$. The proof of the equation of balance in Section 41 is not applicable here, of course, since it is based on a Lorentz system, which need not exist in the Minkowskian manifold $\mathscr{E}$.

To verify (55.2) in the general theory of relativity, we take the covariant derivative of (55.1) converted into the contravariant form,

$$\Omega^{\beta\alpha}{}_{,\beta} = \frac{1}{4\pi}\left(\Sigma_{\mu\nu}\Phi^{\mu\alpha}{}_{,\beta}\Phi^{\nu\beta} + \Sigma_{\mu\nu}\Phi^{\mu\alpha}\Phi^{\nu\beta}{}_{,\beta} - \frac{1}{2}\Sigma^{\alpha\beta}\Phi_{\mu\nu,\beta}\Phi^{\mu\nu}\right). \tag{55.4}$$

Using the field equation (51.23a), we have

$$\xi^{\alpha} = \Phi^{\alpha\beta}J_{\beta} = \frac{1}{4\pi}\Phi^{\alpha\beta}\Phi_{\beta}{}^{\lambda}{}_{,\lambda} = -\frac{1}{4\pi}\Sigma_{\mu\nu}\Phi^{\mu\alpha}\Phi^{\nu\beta}{}_{,\beta}. \tag{55.5}$$

Adding (55.4) and (55.5), we obtain

$$\begin{aligned}\Omega^{\alpha\beta}{}_{,\beta} + \xi^{\alpha} &= \frac{1}{4\pi}\left(\Sigma_{\mu\nu}\Phi^{\mu\alpha}{}_{,\beta}\Phi^{\nu\beta} - \frac{1}{2}\Sigma^{\alpha\beta}\Phi_{\mu\nu,\beta}\Phi^{\mu\nu}\right)\\ &= \frac{1}{8\pi}\Sigma^{\alpha\beta}\Phi^{\mu\nu}(\Phi_{\mu\beta,\nu} + \Phi_{\beta\nu,\mu} + \Phi_{\nu\mu,\beta}).\end{aligned} \tag{55.6}$$

Now from the symmetry of the Christoffel symbols $\{{}^{\alpha}_{\mu\nu}\}$ and the skew-symmetry of $\Phi_{\mu\nu}$ with respect to the pair (μ, ν), we get the identity

$$\Phi_{\mu\beta,\nu} + \Phi_{\beta\nu,\mu} + \Phi_{\nu\mu,\beta} = \frac{\partial\Phi_{\mu\beta}}{\partial x^{\nu}} + \frac{\partial\Phi_{\beta\nu}}{\partial x^{\mu}} + \frac{\partial\Phi_{\nu\mu}}{\partial x^{\beta}}. \tag{55.7}$$

Hence the right-hand side of (55.6) vanishes by virtue of the field equation (51.15a). Thus the equation of balance (55.2) is proved in the general theory.

It follows from (55.2) that $\mathbf{\Omega}$ satisfies the field equations

$$\Omega^{\alpha\beta}{}_{,\beta} = 0 \tag{55.8}$$

in any region where there is no charge–current field $\mathbf{\Psi}$. In the view of Einstein if a field $\mathbf{\Omega}$ of electromagnetic stress-energy-momentum tensors is present in a region, then the field $\mathbf{\Theta}$ of stress-energy-momentum tensors is given by the formula

$$\mathbf{\Theta} = \hat{\varrho}\dot{\mathbf{w}} \otimes \dot{\mathbf{w}} - \mathbf{T} + \mathbf{\Omega}; \tag{55.9}$$

cf. (45.26). In particular when there is no material in the region, the field equations (46.6) reduce to

$$G_{\alpha\beta} = -K\Omega_{\alpha\beta}, \tag{55.10}$$

where K is given by (46.30).

Note. Unlike the contribution of a material medium to $\mathbf{\Theta}$, the contribution of an electromagnetic field to $\mathbf{\Theta}$ is not confined to some small bounded regions in $\mathscr{E}$, since $\mathbf{\Phi}$, and therefore $\mathbf{\Omega}$, differ from zero both inside and outside the domain of the charge–current field $\mathbf{\Psi}$. In other words, in the entire domain of the electromagnetic field $\mathbf{\Phi}$ there is a field $\mathbf{\Omega}$ of electromagnetic stress-energy-momentum tensors, and that field interacts with the gravitational potential (i.e., the Minkowskian metric) through Einstein's field equations.

In the view of Lorentz the electric charge–current field $\mathbf{\Psi}$, like the proper mass distribution ϱ, is confined to certain bounded regions in $\mathscr{E}$. Since, physically, any particle which carries some electric charge also possesses some proper mass, the field $\mathbf{\Theta}$ of stress-energy-momentum tensors in these bounded regions is given by the complicated general formula (55.9). Outside these bounded regions, however, only the electromagnetic field $\mathbf{\Phi}$ is present, and the field equations reduce to (55.10). When we prescribe certain boundary conditions, these equations may be solved to determine the gravitation potential σ, which is due to both the charge and the proper mass in the bounded regions. In the following section we shall summarize a rigorous solution of Einstein's field equations and Maxwell's equations outside a small region occupied by a single electrically charged mass point. As in the Schwarzschild solution the Einstein tensor associated with this solution becomes singular at certain points inside the small region.

Taking the trace of (55.1), we obtain

$$\Sigma^{\alpha\beta}\Omega_{\alpha\beta} = \Omega = 0. \tag{55.11}$$

The preceding result may be compared with the result that the trace of the intrinsic stress-energy-momentum tensor $\varrho\dot{\mathbf{w}} \otimes \dot{\mathbf{w}}$ of a material medium is $-\varrho c^2$, which is the negative of the energy density of the proper mass. Hence (55.11) may be regarded as the assertion that there is no proper mass associated with an electromagnetic stress-energy-momentum tensor. For most cases the contribution of $\mathbf{\Omega}$ to $\mathbf{\Theta}$ is very small compared with that of a material medium. Using (55.11) and (46.12a), we can rewrite the field equations (55.10) as

$$S_{\alpha\beta} = -K\Omega_{\alpha\beta}, \tag{55.12}$$

where $\mathbf{S}$ denotes the Ricci tensor of the Minkowskian metric σ as before.

In Section 48 we formulated an action principle for the gravitational potential σ in a domain which is free of electromagnetic fields; now we generalize that principle to include electromagnetic fields. We define the electromagnetic action integral $\mathsf{E}(\tilde{\sigma}, \tilde{\mathbf{\Pi}})$ on a domain $\mathscr{D}$ in $\mathscr{E}$ by

$$\mathsf{E}(\tilde{\sigma}, \tilde{\mathbf{\Pi}}) \equiv \frac{1}{4\pi}\int_{\mathscr{D}} \tilde{\sigma}_2(\tilde{\mathbf{\Phi}}, \tilde{\mathbf{\Phi}})\tilde{\mathbf{\Xi}} = \frac{1}{8\pi}\int_{\mathscr{D}} \tilde{\Phi}^{\mu\nu}\tilde{\Phi}_{\mu\nu}(-\tilde{\Sigma})^{1/2}\, dw, \tag{55.13}$$

where $dw = dx^1\,dx^2\,dx^3\,dx^4$ as before, and where $\tilde{\Phi}_{\alpha\beta}$ is given by

$$\tilde{\Phi}_{\mu\nu} = \frac{\partial \tilde{\Pi}_\nu}{\partial x^\mu} - \frac{\partial \tilde{\Pi}_\mu}{\partial x^\nu}: \tag{55.14}$$

cf. (51.15b). As explained before, we require that the boundary values of $\tilde{\sigma}$ and $\tilde{\mathbf{\Pi}}$ be held fixed. In other words we consider the variation of the integral $E(\tilde{\sigma}, \tilde{\mathbf{\Pi}})$ over the class of Minkowskian metrics $\tilde{\sigma}$ and electromagnetic potentials $\tilde{\mathbf{\Pi}}$ having the same boundary values on $\partial\mathscr{D}$. Under the preceding assumption we calculate the variation $\delta E(\sigma, \mathbf{\Pi})$ of E at the pair $(\sigma, \mathbf{\Pi})$ in a standard way:

$$\delta E(\sigma, \mathbf{\Pi}) = \frac{1}{8\pi} \int_{\mathscr{D}} \{\Phi_{\alpha\beta}\Phi_{\mu\nu}\delta[\Sigma^{\alpha\mu}\Sigma^{\beta\nu}(-\Sigma)^{1/2}] + 2\Sigma^{\alpha\mu}\Sigma^{\beta\nu}(-\Sigma)^{1/2}\Phi_{\alpha\beta}\delta\Phi_{\mu\nu}\}\, dw. \tag{55.15}$$

As usual the variation $\delta[\Sigma^{\alpha\mu}\Sigma^{\beta\nu}(-\Sigma)^{1/2}]$ is given by

$$\delta[\Sigma^{\alpha\mu}\Sigma^{\beta\nu}(-\Sigma)^{1/2}] = \Sigma^{\beta\nu}(-\Sigma)^{1/2}\delta\Sigma^{\alpha\mu} + \Sigma^{\alpha\mu}(-\Sigma)^{1/2}(\delta\Sigma^{\beta\nu} - \tfrac{1}{2}\Sigma^{\beta\nu}\Sigma_{\gamma\lambda}\delta\Sigma^{\gamma\lambda}), \tag{55.16}$$

where we have used the formula (48.31). Substituting (55.16) into the first term of the integrand in (55.15), we get

$$\begin{aligned}\Phi_{\alpha\beta}\Phi_{\mu\nu}\delta[\Sigma^{\alpha\mu}\Sigma^{\beta\nu}(-\Sigma)^{1/2}] &= 2(\Sigma^{\beta\nu}\Phi_{\alpha\beta}\Phi_{\mu\nu} - \tfrac{1}{4}\Sigma_{\alpha\mu}\Phi^{\gamma\lambda}\Phi_{\gamma\lambda})(-\Sigma)^{1/2}\delta\Sigma^{\alpha\mu} \\ &= 8\pi\Omega_{\alpha\beta}(-\Sigma)^{1/2}\delta\Sigma^{\alpha\beta},\end{aligned} \tag{55.17}$$

where we have used (55.1). From (55.14) the variation $\delta\Phi_{\mu\nu}$ is given by

$$\delta\Phi_{\mu\nu} = \frac{\partial}{\partial x^\mu}(\delta\Pi_\nu) - \frac{\partial}{\partial x^\nu}(\delta\Pi_\mu). \tag{55.18}$$

Substituting (55.18) into the second term of the integrand in (55.15), we get

$$\begin{aligned}2\Phi^{\mu\nu}(-\Sigma)^{1/2}\delta\Phi_{\mu\nu} &= 2\Phi^{\mu\nu}(-\Sigma)^{1/2}\left[\frac{\partial}{\partial x^\mu}(\delta\Pi_\nu) - \frac{\partial}{\partial x^\nu}(\delta\Pi_\mu)\right] \\ &= 4\Phi^{\mu\nu}(-\Sigma)^{1/2}\frac{\partial}{\partial x^\mu}(\delta\Pi_\nu) \\ &= 4\frac{\partial}{\partial x^\nu}[\Phi^{\mu\nu}(-\Sigma)^{1/2}\delta\Pi_\nu] - 4\left\{\frac{\partial}{\partial x^\mu}[\Phi^{\mu\nu}(-\Sigma)^{1/2}]\right\}\delta\Pi_\nu \\ &= 4\frac{\partial}{\partial x^\nu}[\Phi^{\mu\nu}(-\Sigma)^{1/2}\delta\Pi_\nu] + 4(-\Sigma)^{1/2}\Phi^{\nu\mu}{}_{,\mu}\delta\Pi_\nu \\ &= 4\frac{\partial}{\partial x^\nu}[\Phi^{\mu\nu}(-\Sigma)^{1/2}\delta\Pi_\nu] + 16\pi(-\Sigma)^{1/2}J^\nu\delta\Pi_\nu,\end{aligned} \tag{55.19}$$

where we have used (51.21) and (51.23).

Now substituting (55.17) and (55.19) into (55.15) and using the divergence theorem together with the boundary conditions, we obtain

$$\delta E(\sigma, \Pi) = \frac{1}{8\pi} \int_{\mathscr{D}} (8\pi \Omega_{\alpha\beta} \delta \Sigma^{\alpha\beta} + 16\pi J^{\nu} \delta \Pi_{\nu})(-\Sigma)^{1/2} \, dw$$
$$+ \frac{4}{8\pi} \int_{\mathscr{D}} \frac{\partial}{\partial x^{\nu}} [\Phi^{\mu\nu}(-\Sigma)^{1/2} \delta \Pi_{\nu}] \, dw$$
$$= \int_{\mathscr{D}} (\Omega_{\alpha\beta} \delta \Sigma^{\alpha\beta} + 2J^{\nu} \delta \Pi_{\nu}) \Xi, \tag{55.20}$$

which implies that

$$\frac{\delta E(\sigma, \mathbf{\Pi})}{\delta \Sigma^{\alpha\beta}} = \Omega_{\alpha\beta}, \tag{55.21a}$$

$$\frac{\delta E(\sigma, \mathbf{\Pi})}{\delta \Pi_{\nu}} = 2J^{\nu}, \tag{55.21b}$$

where the variational derivatives are defined as before by the condition

$$\delta E(\sigma, \mathbf{\Pi}) = \int_{\mathscr{D}} \left[\frac{\delta E(\sigma, \mathbf{\Pi})}{\delta \Sigma^{\alpha\beta}} \delta \Sigma^{\alpha\beta} + \frac{\delta E(\sigma, \mathbf{\Pi})}{\delta \Pi_{\nu}} \delta \Pi_{\nu} \right] \Xi \tag{55.22}$$

for all variations $\delta \Sigma^{\alpha\beta}$ and $\delta \Pi_{\nu}$, which satisfy the boundary conditions.

In Section 48 we have shown that the Einstein tensor $G_{\alpha\beta}$ may be expressed as the variational derivative of the action integral $A(\sigma)$; cf. (48.22) and (48.23). Combining (48.23) with (55.21), we see that the field equations (55.10) are equivalent to

$$\frac{\delta A(\sigma)}{\delta \Sigma^{\alpha\beta}} + K \frac{\delta E(\sigma, \mathbf{\Pi})}{\delta \Sigma^{\alpha\beta}} = 0. \tag{55.23}$$

Similarly, the condition that there is no charge–current field in the domair $\mathscr{D}$, i.e., $\mathbf{\Psi} = \mathbf{0}$ or $\mathbf{J} = \mathbf{0}$, is equivalent to

$$\frac{\delta E(\sigma, \mathbf{\Pi})}{\delta \Pi_{\nu}} = 0. \tag{55.24}$$

From (55.23) and (55.24) we see that the field equations for any domain $\mathscr{D}$ which is free of any material medium and charge–current field (but there may be an electromagnetic field $\mathbf{\Phi}$ in $\mathscr{D}$) may be characterized by the variational principle

$$\delta[A(\sigma) + KE(\sigma, \mathbf{\Pi})] = 0. \tag{55.25}$$

The preceding assertion is known as the *action principle for an electrovac domain.*

In his lectures Toupin[(6)] remarked that a stress-energy-momentum tensor similar to the electromagnetic stress-energy-momentum tensor may be defined on the basis of a "gravitational field" **G**, which we call a coaction field in Section 49. Specifically we define a "gravitational action" $F(\tilde{\sigma}, \tilde{\Upsilon})$ on a domain $\mathscr{D}$ by

$$F(\tilde{\sigma}, \tilde{\Upsilon}) \equiv \int_{\mathscr{D}} \tilde{\sigma}_1(\tilde{\mathbf{G}}, \tilde{\mathbf{G}})\tilde{\boldsymbol{\Xi}} = \int_{\mathscr{D}} \tilde{G}^\mu \tilde{G}_\mu (-\tilde{\Sigma})^{1/2}\, dw, \tag{55.26}$$

where $\tilde{G}_\mu$ is given by

$$\tilde{G}_\mu = \frac{\partial \tilde{\Upsilon}}{\partial x^\mu}; \tag{55.27}$$

cf. (49.5). The preceding definition for $F(\tilde{\sigma}, \tilde{\Upsilon})$ is quite similar to the definition (55.13) for $E(\tilde{\sigma}, \tilde{\mathbf{\Pi}})$. As before we require that the boundary values of $\tilde{\sigma}$ and $\tilde{\Upsilon}$ be held fixed on $\partial\mathscr{D}$.

From (55.26) we calculate the variation $\delta F(\sigma, \Upsilon)$ of F at the pair (σ, Υ) by

$$\delta F(\sigma, \Upsilon) = \int_{\mathscr{D}} \{G_\alpha G_\beta \delta[\Sigma^{\alpha\beta}(-\Sigma)^{1/2}] + 2\Sigma^{\alpha\beta}(-\Sigma)^{1/2} G_\alpha \delta G_\beta\}\, dw. \tag{55.28}$$

The first term in the integrand may be determined as before by the formula (48.31):

$$G_\alpha G_\beta \delta[\Sigma^{\alpha\beta}(-\Sigma)^{1/2}] = (G_\alpha G_\beta - \tfrac{1}{2}\Sigma_{\alpha\beta} G^\mu G_\mu)(-\Sigma)^{1/2} \delta\Sigma^{\alpha\beta}. \tag{55.29}$$

From (55.27) the variation δG_β is given by

$$\delta G_\beta = \frac{\partial}{\partial x^\beta}(\delta\Upsilon). \tag{55.30}$$

Substituting (55.30) into the second term of the integrand in (55.28), we obtain

$$\begin{aligned}
&2(-\Sigma)^{1/2}\Sigma^{\alpha\beta}\frac{\partial \Upsilon}{\partial x^\alpha}\frac{\partial}{\partial x^\beta}(\delta\Upsilon)\\
&= \frac{\partial}{\partial x^\beta}\left[2\Sigma^{\alpha\beta}(-\Sigma)^{1/2}\frac{\partial\Upsilon}{\partial x^\alpha}\delta\Upsilon\right] - 2\frac{\partial}{\partial x^\beta}\left[\Sigma^{\alpha\beta}(-\Sigma)^{1/2}\frac{\partial\Upsilon}{\partial x^\alpha}\right]\delta\Upsilon\\
&= \frac{\partial}{\partial x^\beta}\left[2\Sigma^{\alpha\beta}(-\Sigma)^{1/2}\frac{\partial\Upsilon}{\partial x^\alpha}\delta\Upsilon\right] - 2(-\Sigma)^{1/2}\,\mathrm{Lap}(\Upsilon)\delta\Upsilon\\
&= \frac{\partial}{\partial x^\beta}\left[2\Sigma^{\alpha\beta}(-\Sigma)^{1/2}\frac{\partial\Upsilon}{\partial x^\alpha}\delta\Upsilon\right] + 2M(-\Sigma)^{1/2}\delta\Upsilon,
\end{aligned} \tag{55.31}$$

where we have used (49.10) and (49.14).

(6) R. A. Toupin, "Electricity and electro-magnetics" (see footnote on page 351).

Now we define a "gravitational stress-energy-momentum tensor" $\mathbf{\Lambda}$ associated with a "gravitational field" $\mathbf{G}$ by the component formula

$$\Lambda_{\alpha\beta} = G_\alpha G_\beta - \tfrac{1}{2}\Sigma_{\alpha\beta}G^\mu G_\mu, \tag{55.32}$$

which is quite similar to the component formula (55.1) for the electromagnetic stress-energy-momentum tensor $\mathbf{\Omega}$. Like the field $\mathbf{\Omega}$, the field $\mathbf{\Lambda}$ satisfies an equation of balance of the form

$$\operatorname{Div} \mathbf{\Lambda} + \boldsymbol{\mu} = 0, \tag{55.33}$$

where $\boldsymbol{\mu}$ is given by the component formula

$$\mu^\alpha = G^\alpha M. \tag{55.34}$$

Here M denotes the dual of the action density $\mathbf{A}$, which gives rise to the "gravitational field" $\mathbf{G}$ as explained in Section 49. The equation of balance (55.33) and the component formula (55.34) are comparable to (55.2) and (55.3).

To verify the equation of balance (55.33), we take the covariant derivative of the component formula (55.32) converted into the contravariant form:

$$\Lambda^{\alpha\beta}{}_{,\beta} = G^\alpha{}_{,\beta}G^\beta + G^\alpha G^\beta{}_{,\beta} - \Sigma^{\alpha\beta}G^\mu{}_{,\beta}G_\mu. \tag{55.35}$$

Now using (49.31), we have

$$\mu^\alpha = G^\alpha M = -G^\alpha \frac{1}{(-\Sigma)^{1/2}} \frac{\partial}{\partial x^\beta}[(-\Sigma)^{1/2}G^\beta] = -G^\alpha G^\beta{}_{,\beta}. \tag{55.36}$$

Adding (55.35) and (55.36), we get

$$\Lambda^{\alpha\beta}{}_{,\beta} + \mu^\alpha = G^\alpha{}_{,\beta}G^\beta - \Sigma^{\alpha\beta}G^\mu{}_{,\beta}G_\mu = \Sigma^{\alpha\gamma}G^\mu(G_{\gamma,\mu} - G_{\mu,\gamma}). \tag{55.37}$$

Now by virtue of the symmetry of the Christoffel symbols $\left\{{\alpha \atop \gamma\mu}\right\}$ with respect to the pair (γ, μ),

$$G_{\gamma,\mu} - G_{\mu,\gamma} = \frac{\partial G_\gamma}{\partial x^\mu} - \frac{\partial G_\mu}{\partial x^\gamma} = \frac{\partial}{\partial x^\mu}\left(\frac{\partial \Upsilon}{\partial x^\gamma}\right) - \frac{\partial}{\partial x^\gamma}\left(\frac{\partial \Upsilon}{\partial x^\mu}\right) = 0, \tag{55.38}$$

where we have used the field equation (49.5). Hence the right-hand side of (55.37) vanishes. Thus the equation of balance (55.33) is proved.

Since the action density $\mathbf{A}$, like the charge–current field $\mathbf{\Psi}$, is confined to certain bounded regions in $\mathscr{E}$, outside these regions the equation of balance reduces to

$$\Lambda^{\alpha\beta}{}_{,\beta} = 0, \tag{55.39}$$

which is comparable to (55.8). It should be noted, however, that the field $\mathbf{G}$, like the field $\mathbf{\Phi}$, need not vanish outside the domain of $\mathbf{A}$ and M. Hence there is a nonzero tensor $\mathbf{\Lambda}$ associated with the tensor $\mathbf{G}$ outside the bounded regions also.

Note. A fundamental difference between the tensors $\mathbf{\Lambda}$ and $\mathbf{\Omega}$ is that, while the trace of $\mathbf{\Omega}$ with respect to σ always vanishes [cf. (55.11)], the trace of $\mathbf{\Lambda}$ may not vanish. Indeed, from (55.31)

$$\Sigma^{\alpha\beta}\Lambda_{\alpha\beta} = \Lambda = -G^{\alpha}G_{\alpha} = -\sigma_1(\mathbf{G}, \mathbf{G}). \tag{55.40}$$

Now using the "gravitational stress-energy-momentum tensor" $\Lambda_{\alpha\beta}$, we can rewrite (55.29) as

$$G_{\alpha}G_{\beta}\delta[\Sigma^{\alpha\beta}(-\Sigma)^{1/2}] = \Lambda_{\alpha\beta}(-\Sigma)^{1/2}\delta\Sigma^{\alpha\beta}. \tag{55.41}$$

Substituting (55.41) and (55.31) into (55.28) and using the divergence theorem together with the boundary conditions, we obtain

$$\begin{aligned}\delta F(\sigma, \Upsilon) &= \int_{\mathscr{D}} (\Lambda_{\alpha\beta}\delta\Sigma^{\alpha\beta} + 2M\delta\Upsilon)(-\Sigma)^{1/2}\, dw \\ &\quad + \int_{\mathscr{D}} \frac{\partial}{\partial x^{\beta}}\left[2\Sigma^{\alpha\beta}(-\Sigma)^{1/2}\frac{\partial\Upsilon}{\partial x^{\alpha}}\,\delta\Upsilon\right] dw \\ &= \int_{\mathscr{D}} (\Lambda_{\alpha\beta}\delta\Sigma^{\alpha\beta} + 2M\delta\Upsilon)\mathbf{\Xi},\end{aligned} \tag{55.42}$$

which implies that

$$\frac{\delta F(\sigma, \Upsilon)}{\delta\Sigma^{\alpha\beta}} = \Lambda_{\alpha\beta}, \tag{55.43a}$$

$$\frac{\delta F(\sigma, \Upsilon)}{\delta\Upsilon} = 2M. \tag{55.43b}$$

Recall that the scalar M in (55.43b) is just the dual of the 4-form $\mathbf{A}$. Hence (55.43b) is similar to (55.21b), where the 1-form $\mathbf{J}$ is the dual of the closed 3-form $\mathbf{\Psi}$.

In his lectures Toupin suggested that the action principle (55.25) might be regarded as a special case of the following more general variational principle:

$$\delta[A(\sigma) + KE(\sigma, \mathbf{\Pi}) + HF(\sigma, \Upsilon)] = 0, \tag{55.44}$$

where K and H are certain universal constants. From (48.23), (55.21), and (55.43) the variational principle (55.44) is equivalent to the following system of field equations:

$$G_{\alpha\beta} + K\Omega_{\alpha\beta} + H\Lambda_{\alpha\beta} = 0, \tag{55.45a}$$

$$J_\alpha = 0, \tag{55.45b}$$

$$M = 0. \tag{55.45c}$$

The field equations (55.45a) are more general than the field equations (55.10), since the "gravitational field" **G** may interact with the gravitational potential σ, provided that the constant H is nonzero. Toupin did not make clear in his lectures whether or not the constant H might vanish, however.

56. Electrogravitational Fields of an Electrically Charged Mass Point

In this section we derive a rigorous solution of the coupled system of Einstein's equations (55.10) and Maxwell's equations (51.15) outside a small region occupied by an electrically charged mass point; that solution was obtained originally by Nordström and by Jeffrey.

We start from the assumption that the components of the gravitational potential (i.e., the Minkowskian metric σ) and the electromagnetic potential **Π** are of the forms

$$\Sigma_{11} = e^{\lambda(r)}, \qquad \Sigma_{22} = r^2, \qquad \Sigma_{33} = r^2 \sin^2\theta, \qquad \Sigma_{44} = -e^{\nu(r)}, \tag{56.1}$$

and

$$\Pi_1 = \Pi_2 = \Pi_3 = 0, \qquad \Pi_4 = h(r). \tag{56.2}$$

As remarked in Section 47, the potentials having the special component forms are spherically symmetric with respect to a particular world line, which is characterized by the condition $r = 0$ in the coordinate system $(x^\alpha) = (r, \theta, \varphi, ct)$, where (r, θ, φ) denote the spherical coordinates centered at the world line.

In Section 47 we showed that the nonzero covariant components of the Ricci tensor $\mathbf{S}$ associated with the metric σ are given by

$$\begin{aligned} S_{11} &= \tfrac{1}{2}\nu'' - \tfrac{1}{4}\lambda'\nu' + \tfrac{1}{4}\nu'^2 - \frac{\lambda'}{r}, \\ S_{22} &= e^{-\lambda}[1 + \tfrac{1}{2}r(\nu' - \lambda')] - 1, \\ S_{33} &= e^{-\lambda}\sin^2\theta[1 + \tfrac{1}{2}r(\nu' - \lambda')] - \sin^2\theta, \\ S_{44} &= e^{(\nu-\lambda)}\left[-\frac{\nu''}{2} + \frac{\lambda'\nu'}{4} - \frac{\nu'^2}{4} - \frac{\nu'}{r}\right]; \end{aligned} \tag{56.3}$$

cf. (47.5). Unlike the Ricci tensor of the Schwarzschild solution, the Ricci tensor here does not vanish but is equal to $-K$ times the electromagnetic stress-energy-momentum tensor $\mathbf{\Omega}$, as required by the field equations (55.12). We proceed now to determine the tensor $\mathbf{\Omega}$ from the electromagnetic potential $\mathbf{\Pi}$.

From (56.2) and (51.15b) the nonzero covariant components of the electromagnetic field $\Phi_{\alpha\beta}$ are

$$\Phi_{14} = -\Phi_{41} = \frac{\partial \Pi_4}{\partial x^1} = h'. \tag{56.4}$$

Then the nonzero contravariant components of $\mathbf{\Phi}$ are given by

$$\Phi^{41} = -\Phi^{14} = \Sigma^{11}\Sigma^{44}\Phi_{41} = e^{-(\lambda+\nu)}h', \tag{56.5}$$

since from (56.1) the nonzero components of the dual metric σ_1 are

$$\Sigma^{11} = e^{-\lambda(r)}, \quad \Sigma^{22} = 1/r^2, \quad \Sigma^{33} = 1/r^2\sin^2\theta, \quad \Sigma^{44} = -e^{-\nu(r)} \tag{56.6}$$

From (51.21) the components $\Phi^{\alpha\beta}$ satisfy the field equations

$$\frac{\partial}{\partial x^\beta}[(-\Sigma)^{1/2}\Phi^{\alpha\beta}] = 0 \tag{56.7}$$

outside the small region occupied by the charged mass point. From (56.1) the determinant $\Sigma = \det[\Sigma_{\alpha\beta}]$ is given by

$$\Sigma = -e^{(\lambda+\nu)}r^4\sin^2\theta. \tag{56.8}$$

Substituting (56.5) and (56.8) into (56.7), we obtain

$$\frac{d}{dr}(e^{-(1/2)(\lambda+\nu)}r^2h') = 0. \tag{56.9}$$

Hence

$$h' = \frac{a}{r^2}e^{(1/2)(\lambda+\nu)}, \tag{56.10}$$

where a is a constant of integration. It follows from (56.10), (56.5), and (56.4) that the nonzero components of $\mathbf{\Phi}$ are

$$\Phi_{14} = -\Phi_{41} = \frac{a}{r^2} e^{(1/2)(\lambda+\nu)}, \tag{56.11}$$

and

$$\Phi^{41} = -\Phi^{14} = \frac{a}{r^2} e^{-(1/2)(\lambda+\nu)}. \tag{56.12}$$

Now using the general formula (55.1), we calculate the components of the electromagnetic stress-energy-momentum tensor $\mathbf{\Omega}$, and the result is

$$[\Omega_{\alpha\beta}] = \frac{1}{8\pi} \frac{a^2}{r^4} \operatorname{diag}(-e^{\lambda}, r^2, r^2 \sin^2\theta, e^{\nu}). \tag{56.13}$$

Substituting the component formulas (56.3) and (56.13) into the field equations

$$S_{\alpha\beta} = -\frac{8\pi k}{c^4} \Omega_{\alpha\beta}, \tag{56.14}$$

we obtain the following system of differential equations for the two unknown functions λ and ν:

$$\frac{1}{2}\nu'' - \frac{1}{4}\lambda'\nu' + \frac{1}{4}(\nu')^2 - \frac{\lambda'}{r} = \frac{ka^2}{c^4r^4} e^{\lambda}, \tag{56.15a}$$

$$e^{-\lambda}\left[1 + \frac{1}{2} r(\nu' - \lambda')\right] - 1 = -\frac{ka^2}{c^4r^4} r^2, \tag{56.15b}$$

$$e^{-\lambda}\sin^2\theta\left[1 + \frac{1}{2} r(\nu' - \lambda')\right] - \sin^2\theta = -\frac{ka^2}{c^4r^4} r^2 \sin^2\theta, \tag{56.15c}$$

$$e^{(\nu-\lambda)}\left[-\frac{\nu''}{2} + \frac{\lambda'\nu'}{4} - \frac{(\nu')^2}{4} - \frac{\nu'}{4}\right] = -\frac{ka^2}{c^4r^4} e^{\nu}. \tag{56.15d}$$

This system may be solved easily as in Section 47. [Notice that the homogeneous system of (56.15) is just the system in Section 47 which governs the unknown functions λ and ν in the Schwarzschild solution.]

Specifically, the equations (56.15a) and (56.15d) imply that $\nu' = -\lambda'$. Hence by requiring that, as $r \to \infty$, the metric σ approaches the flat metric $\tilde{\sigma}$ with components

$$\tilde{\Sigma}_{11} = 1, \qquad \tilde{\Sigma}_{22} = r^2, \qquad \tilde{\Sigma}_{33} = r^2 \sin^2\theta, \qquad \tilde{\Sigma}_{44} = -1; \tag{56.16}$$

cf. (47.6), we get $\nu = -\lambda$. From that result the equation (56.15b) reduces to

$$e^{\nu}(1 + r\nu') - 1 = -\frac{ka^2}{c^4r^2}, \tag{56.17}$$

or, equivalently,

$$\frac{d}{dr}(re^{\nu}) = 1 - \frac{ka^2}{c^4r^2}. \tag{56.18}$$

Integrating the preceding equation, we obtain

$$e^{\nu} = 1 - \frac{2km}{c^2r} + \frac{ka^2}{c^4r^2}, \tag{56.19}$$

where m is another constant of integration. It follows from (56.19), (56.6), and (56.1) that the nonzero components of the metric σ are

$$\begin{aligned} &\Sigma_{11} = 1\Big/\left(1 - \frac{2km}{c^2r} + \frac{ka^2}{c^4r^2}\right), && \Sigma_{22} = r^2, \\ &\Sigma_{33} = r^2 \sin^2\theta, && \Sigma_{44} = -\left(1 - \frac{2km}{c^2r} + \frac{ka^2}{c^4r^2}\right), \end{aligned} \tag{56.20}$$

and

$$\begin{aligned} &\Sigma^{11} = 1 - \frac{2km}{c^2r} + \frac{ka^2}{c^4r^2}, && \Sigma^{22} = 1/r^2, \\ &\Sigma^{33} = 1/r^2 \sin^2\theta, && \Sigma^{44} = -1\Big/\left(1 - \frac{2km}{c^2r} + \frac{ka^2}{c^4r^2}\right). \end{aligned} \tag{56.21}$$

Now applying the approximate formula (46.8) to the preceding exact solution, we see that the metric σ corresponds to a Newtonian gravitational potential ζ of the form

$$\zeta = -\frac{1}{2}c^2F_{44} = -\frac{1}{2}c^2\left(\frac{2km}{c^2r} - \frac{ka^2}{c^4r^2}\right) = -\frac{km}{r^2} + \frac{ka^2}{2c^2r^2}. \tag{56.22}$$

We recognize immediately that the leading term on the right-hand side of (56.22) corresponds to the gravitational potential in the classical theory due to a mass point with mass m located at the center $r = 0$ of the spherical coordinate system (r, θ, φ). The second term on the right-hand side of (56.22) is a relativistic correction of the classical formula due to the electric charge of the mass point. Because of the factor k/c^2 in the second term, the correction is very small compared to the leading term.

The reason that the second term on the right-hand side of (56.22) is due to the electric charge may be seen in the following way: Since $\lambda + \nu = 0$, the formulas (56.11) and (56.12) reduce to

$$\Phi_{14} = -\Phi_{41} = \Phi^{41} = -\Phi^{14} = \frac{a}{r^2}. \tag{56.23}$$

We recognize immediately that these components are just the components of the electric field **E** according to Coulomb's law in the classical theory, provided that we regard the constant a as the electric charge of the mass point.

From (56.20) and (56.21) we see that the Nordström–Jeffrey solution, like the classical solution and the Schwarzschild solution, becomes singular at the center $r = 0$; the presence of other singularities like the Schwarzschild singularity depends on the value of the electric charge a relative to the value of the proper mass m. There are three possibilities:

(i) $a^2 > km^2$. In this case $r = 0$ is the only singularity of the solution. The empirical values of a and m for a single electron are approximately

$$a = -4.801 \times 10^{-10} \text{ esu}, \qquad m = 9.109 \times 10^{-28} \text{ g}, \tag{56.24}$$

which satisfy the hypothesis $a^2 > km^2$. For the values given by (56.24) the two terms on the right-hand side of (56.22) are equal in magnitude at the radius

$$r_0 = \frac{a^2}{2c^2m} = 1.41 \times 10^{-13} \text{ cm}, \tag{56.25}$$

which is generally agreed to be of the order of magnitude of the radius of an electron. Differentiating (56.22) with respect to r, we see that ζ has a minimum value at $r = 2r_0$. When $r < 2r_0$, the second term on the right-hand side of (56.22) dominates the first term, and $\zeta \to +\infty$ as $r \to 0$; the form of ζ in the region is entirely different from that of the Schwarzschild solution. On the other hand, when $r > 2r_0$, the first term dominates the second term, and $\zeta \to 0$ as $r \to \infty$; the form of ζ in the region is similar to that of the Schwarzschild solution as well as that of the classical Newtonian solution.

(ii) $a^2 = km^2$. In this case other than the singularity at $r = 0$ there is another singularity in the Nordström–Jeffrey solution at

$$r = \frac{km}{c^2}, \tag{56.26}$$

which is exactly one-half of the radius of the Schwarzschild singularity corresponding to the mass point without the electric charge.

(iii) $a^2 < km^2$. In this case the algebraic equation

$$r^2 - \frac{2km}{c^2} r + \frac{ka^2}{c^4} = 0 \tag{56.27}$$

has two positive roots:

$$r = \frac{km}{c^2} \left[1 \pm \left(1 - \frac{a^2}{km^2} \right)^{1/2} \right], \tag{56.28}$$

which are both less than $2km/c^2$, the radius of the Schwarzschild singularity corresponding to the mass point without the electric charge. [In fact, from (56.27) the sum of the two positive roots given by (56.28) is equal to $2km/c^2$. The radius given by (56.26) in the previous case corresponds to a double root of the equation (56.27) when the coefficients satisfy the condition $a^2 = km^2$.] When a^2 is much smaller than km^2, i.e., when the contribution of the electric charge is small compared to that of the proper mass, the two roots are given approximately by

$$r_1 = \frac{2km}{c^2} - \frac{a^2}{2mc^2}, \qquad r_2 = \frac{a^2}{2mc^2}, \tag{56.29}$$

where one root, r_1, is close to $2km/c^2$, while the other root, r_2, is close to the center. Since the radius $2km/c^2$ of the Schwarzschild singularity of the corresponding mass point without the electric charge is an upper bound for the two roots (56.28), by the remark made in Section 47 we cannot observe these singularities outside the region occupied by the charged mass point.

Selected Reading for Part B

BERGMANN, P., *Introduction to the Theory of Relativity*, Prentice-Hall, Englewood Cliffs, New Jersey (1942).

BERGMANN, P., The special theory of relativity, in Flügge's *Handbuch der Physik*, Band IV, Springer, Berlin, 1962.

BERGMANN, P., The General Theory of Relativity, in Flügge's *Handbuch der Physik*, Band IV, Springer, Berlin, 1962.

BOWEN, R. M. and C.-C. WANG, *Introduction to Vectors and Tensors*, Volumes 1 and 2, Plenum, New York, 1976.

EDDINGTON, A. S., *The Mathematical Theory of Relativity*, Cambridge University Press, Cambridge, England, 1954.

FEYNMAN, R. P., *Lectures on Physics*, Addison-Wesley, Reading, Massachusetts, 1964.

FLANDERS, H., *Differential Forms*, Academic Press, New York, 1963.

HODGE, W. V. D., *The Theory and Applications of Harmonic Integrals*, Cambridge University Press, Cambridge, England, 1952.

LANDAU, L. and E. LIFSCHITZ, *Electrodynamics of Continuous Media*, Course of Theoretical Physics, Volume 8, Pergamon Press, London, 1960.

PHILLIPS, M., Classical electrodynamics, in Flügge's *Handbuch der Physik*, Band IV, Springer, Berlin, 1962.

SCHOUTEN, J. A., *Ricci Calculus*, Springer, Berlin, 1954.

SOMMERFELD, A., *Electrodynamics*, Lectures on Theoretical Physics, Volume 3, Academic Press, New York, 1952.

STRATTON, J. A., *Electromagnetic Theory*, McGraw-Hill, New York, 1941.

SYNGE, J. L., *Relativity: The Special Theory*, North-Holland, Amsterdam, 1956.

Synge, J. L., *Relativity: The General Theory*, North-Holland, Amsterdam, 1960.

Toupin, R. A., Elasticity and electro-magnetics, in *Non-linear Continuum Theories*, pp. 206–342, C.I.M.E. lecture notes, Bressanone, Italy, 1965; conference coordinators: C. Truesdell and G. Grioli.

Whitney, H., *Geometric Integration Theory*, Princeton University Press, Princeton, New Jersey, 1957.

Whittaker, E., *A History of the Theories of Aether and Electricity* (two volumes), Nelson, London, 1951, 1953.

Index

Pages 1–198 will be found in Part A, pages 199–386 in Part B.